THIS PLACE ON EARTH 2002

OTHER NORTHWEST ENVIRONMENT WATCH TITLES

This Place on Earth 2001: Guide to a Sustainable Northwest

State of the Northwest, revised 2000 edition

Seven Wonders: Everyday Things for a Healthier Planet

Green-Collar Jobs

Tax Shift

Over Our Heads: A Local Look at Global Climate

Misplaced Blame: The Real Roots of Population Growth

Stuff: The Secret Lives of Everyday Things

This Place on Earth: Home and the Practice of Permanence

The Car and the City

Hazardous Handouts: Taxpayer Subsidies to Environmental Degradation

State of the Northwest

THIS PLACE ON EARTH 2002

Measuring What Matters

NORTHWEST ENVIRONMENT WATCH ◆ SEATTLE

NORTHWEST ENVIRONMENT WATCH IS AN INDEPENDENT, not-for-profit research and communication center in Seattle, Washington. Its mission is to promote a sustainable economy and way of life throughout the Pacific Northwest—the biological region stretching from southeast Alaska to northern California and from the Pacific Ocean to the crest of the Rockies.

Library of Congress Control Number: 2002100162
ISBN 1-886093-12-1

Cover illustration: Sooah Lee
Cover and interior design: Cathy Schwartz
Interior illustration: Sooah Lee (page ix), Rebecca Wayt (page 1), Diana Moore (page 51), and Cathy Schwartz (all graphs); base maps by CommEn Space, Seattle (pages 27–32), and Dorie Brownell, Ecotrust, Portland (page 33)
Editing and composition, including maps: Ellen W. Chu
Proofreading: Sherri Schultz

Printed by Artcraft Printing Company, Seattle, Washington, with vegetable-based ink on recycled paper. Text: 100 percent postconsumer waste, bleached without chlorine; map pages: 35 percent preconsumer and 15 percent postconsumer waste, bleached without chlorine. Cover: 50 percent preconsumer waste, bleached with hydrogen peroxide.

Northwest Environment Watch is a 501(c)(3) tax-exempt organization. To order publications, become a member, or learn more, please contact:

Northwest Environment Watch
1402 Third Avenue, Suite 500
Seattle, WA 98101-2130 USA
(206) 447-1880; fax (206) 447-2270
new@northwestwatch.org; www.northwestwatch.org

CONTENTS

ACKNOWLEDGMENTS

This Place on Earth 2002 was researched and written by Alan Thein Durning, Clark Williams-Derry, Eric de Place, and Ellen W. Chu, with additional research by former research associate Joanna Lemly and past research interns. For their helpful comments, Northwest Environment Watch (NEW) thanks reviewers Alan AtKisson, Stephen Bezruchka, Daniel Bottom, Ralph Cipriani, Robert Francis, Cheeying Ho, Robert Hughes, James Karr, Aaron Katz, Jim Kerstetter, John Kramlich, Jim Lazar, Conway Leovy, Jim Lichatowich, Fiona MacPhail, Philip Malte, Nathan Mantua, Langdon Marsh, Richard Morrill, Phillip Mote, Charles Treser, John M. Wallace, and Carolyn Watts.

Students at Seattle's Cornish College of the Arts created the cover and inside paintings as a class project for instructor Bonnie Rieser. Many thanks to Sooah Lee, Diana Moore, and Rebecca Wayt for their patience and hard work in illustrating "a sustainable Northwest."

NEW also thanks its volunteers Jeffrey Belt, Dan Bertolet, Bill Kint, Norman Kunkel, Greg Martin, Lyn McCollum, Ashley Mitchell, Michael Montague, Neal Parry, Marcus Pina, Tim Varga, John Wedgwood, Erin Weible, Suzy Whitehead, and Tony Zamparutti. For revamping our Web site, special thanks go to Steve Mack.

NEW's financial support comes from its members and donors, and from the Brainerd and Bullitt Foundations; the Charitable Gift Fund; the C. S. Fund; and the Contorer, Nathan Cummings, Glaser, Hanlon, William and Flora Hewlett, Kongsgaard-Goldman, Lazar, Mountaineers, David and Lucile Packard, Russell Family, True North, Weeden, and Winky Foundations.

In 2001 NEW benefited from the generosity of those who hosted or headlined events on its behalf: Sherman Alexie, John and Elizabeth Atcheson, Aaron Contorer, John and Jane Emrick, Mike and Molly Hanlon, Paul Hawken, Bill McKibben, Drummond Pike and the Tides Foundation, Jonathan Raban, Laura Retzler and Henry Wigglesworth, John Russell, Joan Sawicki, Kathy Fong Stephens, and David Yaden.

Finally, NEW is grateful to its board of directors: David Yaden, chair, and John Atcheson, Aaron Contorer, Alan Durning, Jeff Hallberg, Sandra Hernshaw (term ended in 2001), Cheeying Ho, and Laura Retzler (term began in 2001).

NEW's staff while *This Place on Earth 2002* was in progress comprised Alan Thein Durning, executive director; Parke G. Burgess, managing director; Ellen W. Chu, editorial director; Rhea Connors, operations director; Eric de Place and Joanna Lemly, research associates; Michelle Hoar, BC communications associate; Tyesha Kobel, program coordinator; Elisa Murray, communications director; Stacey Panek, membership coordinator; Leigh Sims and Chad Westmacott, communications associates; Deirdre Stevenson, bookkeeper; and Clark Williams-Derry, research director.

THIS PLACE

LOOK DOWN. ABOUT TWO MILES BENEATH your feet, living organisms peter out. Look up. About six miles overhead, the same thing happens. Between these boundaries, life fills every conceivable niche; beyond them, complex life ceases. Microbes may inhabit other planets; they may even be commonplace. But the odds are overwhelmingly stacked against animal life, and life like humans—with language, culture, science, religion—is an infinitesimal chance, a miracle. Statistically, we should not have happened.[1]

Look around you. Odds are, you can see evidence of other people. We humans have adapted to virtually all terrestrial habitats and generated more living matter at one time than any other species ever. In the last 300 years, we have increased our average body size by half; doubled our average lifespan; multiplied our number tenfold; and increased our use of water, soils, plants, and animals many hundredfold. But this success has jeopardized much else on Earth, destabilizing the global climate and dooming species to extinction at a rate exceeding one per hour. One of this century's defining challenges is to achieve sustainability: an economy and way of life in which both people and nature are thriving.[2]

In this global quest, the Pacific Northwest (frontispiece) stands out as a promising test case: The region has 16 million human residents, a $450 billion economy, and a larger share of its ecosystems intact than perhaps any other part of the industrial world. The region has traditions of innovation in business and government; an educated populace; and a longstanding commitment to conservation. If northwesterners can reconcile themselves with the landscapes of this place, they will set an example for the world.[3]

GROWTH: *of* WHAT, *for* WHAT?

IN THE IMMEDIATE AFTERMATH OF THE September 11 terrorist attacks, even in the grip of the world's rage and grief, one of the media's most persistent questions was, what was the death toll? It seems a peculiar thing to ask. Each individual death was a tragedy incommensurable with any other. For the victims' families, a precise body count could change nothing. Still, the question, how many deaths? kept coming up as people searched for a frame of reference, for some quantification of the incomprehensible. The question is profoundly revealing. Counting things is a deeply human impulse; it's one way we understand our world.[4]

This book is about what our society counts in its attempt to assess itself and make choices for the future. From deaths to births, from economic growth to standardized test scores, measurement permeates contemporary life. Hundreds of measurements—or indicators—fill the news, shape public opinion, and inform the millions of actions that individuals and organizations take each day. They serve as proxies for larger, more complicated trends, telling us whether the state of the human enterprise is getting better or worse.

But many of the indicators we rely on, particularly economic ones such as the Dow Jones industrial average, are deeply flawed. They conceal what they purport to reveal and, in the process, systematically misinform us, misdirecting our actions on a grand scale. *This Place on Earth 2002* takes a step toward measuring what really matters to northwesterners. This book documents ten gauges of whether people and nature in the Pacific Northwest are secure and thriving. These measures are promising but unfinished; their reliability compares favorably with society's prevailing indicators, but only because those indicators are deeply flawed.

The Dow Jones industrial average, for example, is the undisputed king of stock market yardsticks. In popular consciousness, the Dow *is* the market; the NASDAQ composite, S & P 500, and other measures merely color the news. But the Dow is the least-accurate regularly published stock indicator. Not only is its roster of 30 companies assembled without any particular methodology—two people at Dow Jones & Co. simply pick them—but the index is also mathematically spurious. The Dow averages stock prices while ignoring the companies' total value.[5]

The Dow's odd structure is explained by its genesis. Charles Dow, first editor of the *Wall Street Journal,* created it in 1896, scribbling out the calculations by hand. A short list of stocks and a simple formula let him report his average easily and often. Incorporating market capitalization (the number of shares times the price of each) would have necessitated a lot of multiplying and adding, followed by one gigantic division problem. The world's most quoted economic gauge may therefore be erroneous because of a fear of long division.[6]

If the Dow measures badly, another leading indicator, gross domestic product (GDP), measures well the wrong thing—or rather, it does not measure what we think it does. North Americans take GDP as the bellwether of national well-being, but GDP doesn't track how people are, only how much they spend.[7]

GDP fails to distinguish between losses and gains, because it only adds and doesn't subtract. Gutting ecosystems for commodities—and leaving fisheries depleted, forests cleared, or rivers dammed—shows up as a plus in the accounts. So do expensive misfortunes: whether money is spent on vacations or hospital stays, playground equipment or car wrecks, births or funerals, it's all the same in the GDP ledger. Likewise, GDP goes up regardless of whether consumers may regret the purchase (such as spending on alcohol, tobacco, or gambling) or the need for the purchase (such as spending on car alarms, firearms, gated communities, or private guards). Blind to services provided free by families, friends, and communities, GDP math values an hour of paid daycare more highly than an hour of unpaid parenting and a book from a store more than one from a library.[8]

With its blinkered accounting, GDP cannot see much of what is important. It tells only about growth in economic production, not in economic satisfaction or quality of life. Since 1957, for example, US real GDP per capita has more than doubled, but the share of Americans who describe themselves as "very happy" has remained unchanged at about one-third. Even Simon Kuznets, Nobel Prize–winning economist and inventor of the GDP, warned against treating it as a gauge of progress; in 1962, he wrote, "Goals for 'more' growth should specify more growth of what and for what."[9]

The world's most quoted index may owe its structure to a fear of long division

Consider the events of 2001 as they looked on the GDP ledger: February's earthquake in the Puget Sound region looked like a boom in building repairs. The extended spring strike that shut down transit—and increased driving—in Vancouver, BC, showed up as a jump in gasoline spending. The mysterious plant disease that swept up the coast

irtime america's best colleges and universities american college test scores american top 40 from California to Oregon, killing the states' revered oaks and threatening its redwoods, looked like a banner year for arborists and tree nurseries. In eastern Oregon and Washington, the catastrophic wildfires of an exceptionally dry summer appeared as huge outlays for public and private firefighting. Then, in September, the World Trade Center attack showed up as a surge in sales of flags, flowers, and relief packages sent East, plus a sudden slackening of spending on air travel and tourism and, later, of spending generally.

Just as the Dow and GDP deceive us, other indicators also give inaccurate readings. The consumer price index exaggerates inflation (because part of its rise reflects tastes that change with purchasing power); the unemployment rate understates unemployment (because it excludes those who have given up on finding jobs); and the poverty rate undercounts the poor (because the US poverty definition is the most miserly standard of basic needs in the industrial world).[10]

The Northwest's indicators should spring from the region's values, not just from the marketplace

What society measures is sometimes a fluke of history, a canonization of the oft repeated. That the Dow's every vacillation is reported in excruciating detail by media outlets worldwide is a testament to the self-reinforcing nature of public perceptions. Sometimes society simply relies on indicators that are easy to count or that businesses or public agencies are tallying for their own reasons anyway: hence such regular reports as payroll, sales tax receipts, unemployment insurance filings, profits, commodity production, births, deaths, and housing starts. But often, what gets counted flows from political power. Businesses want to know buyers' moods, so business associations measure consumer confidence; governments deploy dozens of workers to monitor inflation because the financial and labor sectors want to know.

The Northwest's chosen indicators should not be historical flukes, matters of convenience, or perquisites of power. They should spring from the region's values, its aspirations for the future. Financial security ranks high among those values, so it is fitting that the Northwest monitor its financial capital. But it is not fitting that financial measurements should overwhelm all others in how often they are tabulated; that stock quotes, commodity prices, and other yardsticks of the marketplace should crowd out indicators of community vitality, human well-being, and ecological integrity.[11]

Ideally, indicators of the Northwest's lasting progress—its sustainability—would measure to what extent northwesterners are secure and thriving, to what extent Northwest nature is thriving, and to what extent northwesterners' way of life is benign in its impacts on nature and cultures outside the region. Of course, *secure* and *thriving* leave much undefined. For northwesterners, the words connote mental and physical health; safety from violence; decent livelihoods; longevity; and, as important, a life full of such intangibles as friendship, love, respect, community, a sense of purpose, autonomy, and opportunities to develop interests and talents—plus justice, freedom, and democracy. For nature to be thriving, native plants and animals and the ecological and evolutionary processes that sustain them must persist, spared from disruption or serious degradation at human hands.

As a practical matter, measuring whether people and nature are thriving is difficult. *This Place on Earth 2002* moves toward that end by examining an eclectic set of ten indicators, each critical in a different way to human quality of life, ecosystem health, or both. These ten, which Northwest Environment Watch (NEW) chose because they seemed promising and because data were available, make a start but remain unsatisfactory in a number of ways. They do not go far enough to measure human well-being (among the indicators that follow, only "Health" and "Income" do so directly) or ecosystem health (only

"Salmon" does). Most of the others measure impacts on the environment or the efficiency with which northwesterners use natural resources.

Still, the preliminary set of indicators in this book does allow a peek behind the headlines at some of the lasting changes sweeping the region over the past months and years:

- Life expectancy increased gradually throughout the region, signaling that northwesterners' health improved modestly. Babies born in 2001 are expected to live almost two months longer than babies born a year earlier.
- The fruits of the Northwest's prosperity continued to accrue disproportionately to those at the top of the income ladder, even as the region skidded to the end of a 20-year economic boom. More northwesterners lived in poverty in 2001 than in 1981.
- The population of the Northwest grew slightly more slowly in 2001 than in 2000 and much more slowly than in the early 1990s, though it still added an estimated 22 new northwesterners each hour.
- One in three residents of the Northwest's great metropolitan areas lives in a compact, "smart growth" community after a decade of rapid expansion. Seattle grew outward in the 1990s, sprawling across the landscape. Vancouver, BC, grew denser, as newcomers moved into compact neighborhoods where transit and walking are viable alternatives to driving. Portland split the difference between Seattle and Vancouver.
- Despite stepped-up efforts to stem it, sprawl continued to overrun rural land and commit northwesterners to a car-centered lifestyle. In greater Portland, Seattle, and Vancouver, BC, development's footprint expanded by an average of 13 square miles (34 square kilometers) a year during the 1990s.

- Motor vehicles proliferated at the same pace as population through the 1990s, having outpaced it previously. But as people switched from cars to trucks, vehicles grew bigger and less fuel-efficient.
- After three decades of relentless growth, the region's network of inventoried roads ceased to expand in the 1990s, as old national forest roads in the Northwest states were obliterated as fast as new paved roads and highways were built across the region.
- Although returning salmon runs in 2001 were among the largest in years, the numbers came nowhere near historical averages, and scores of unique, locally adapted stocks—cultural totems and prime indicators of healthy ecosystems—remain imperiled.
- Despite recent efficiency gains, population growth continued to push up regional energy consumption.
- The Northwest's releases of greenhouse gases into the atmosphere, long on the increase, remained steady in 2001. Reduced emissions from aluminum smelters, which closed because of hydropower shortages, were offset by growing releases from fossil fuels.

One in three of the Northwest's major metropolitan residents lives in a "smart growth" community

One unwelcome pattern emerging from the swirl of change is the widening disparity between the life prospects for northwesterners who have marketable skills and those who do not. The Northwest is becoming a bifurcated society: The fortunate enjoy a lifestyle of rising incomes and unparalleled recreational opportunities, epitomized by the region's proliferating resorts, ski chalets, and second homes. The unfortunate struggle to avoid hunger; live in overcrowded housing, often in trailer parks; or, all too often, end up in prison—the fastest-growing form of housing in the Pacific Northwest during the past decade (see Figure 1).[12]

Another, more welcome pattern is that, per person, environmental impacts, at least those we measure, have stopped getting bigger. Vehicle

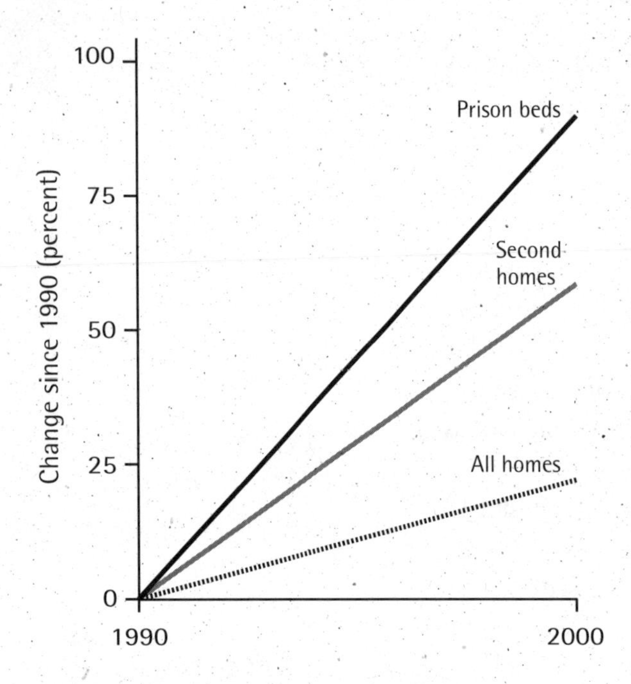

Figure 1. Prisons were the fastest-growing form of housing in the Northwest states in the 1990s, followed by second homes. Sources: see endnote 12.

numbers, energy use, and greenhouse gas emissions have unhitched themselves from the rate of income growth and now rise, more slowly, in rough proportion to the number of northwesterners (see Figure 2). Unfortunately, however, per capita environmental impacts are still high, and population keeps growing.[13]

In aggregate, the scale of human disruption of natural systems goes far beyond what nature can sustain, so slowed growth of that disruption is only a first step toward eventual reconciliation. Stabilizing global climate, for example, will require the Northwest to reduce greenhouse gas emissions by 80 to 90 percent, a change possible only through profound reordering of the region's ways with energy, transportation, forestry, and settlement. Restoring the region's wild salmon runs—many of which are now at less than 10 percent of their historical abundance and diversity—entails similarly sweeping change in many sectors of the economy and spheres of life.[14]

The opportunity at hand for the Pacific Northwest is to apply its talented pragmatism—the stuff that has made the region an entrepreneurial hotspot—to the challenge of shrinking the regional economy's ecological footprint by an order of magnitude. The danger is that the Northwest will continue to split in two, dividing by class into separate and increasingly antagonistic enclaves.

But until the Northwest begins measuring what it values, instead of valuing what it measures, it will not be able to seize the opportunity or avoid the danger. When, however, the region does regularly monitor its environmental and social performance along with its economy, this place on Earth may yet achieve a way of life that can last—one that nurtures human community while honoring nature's limits.

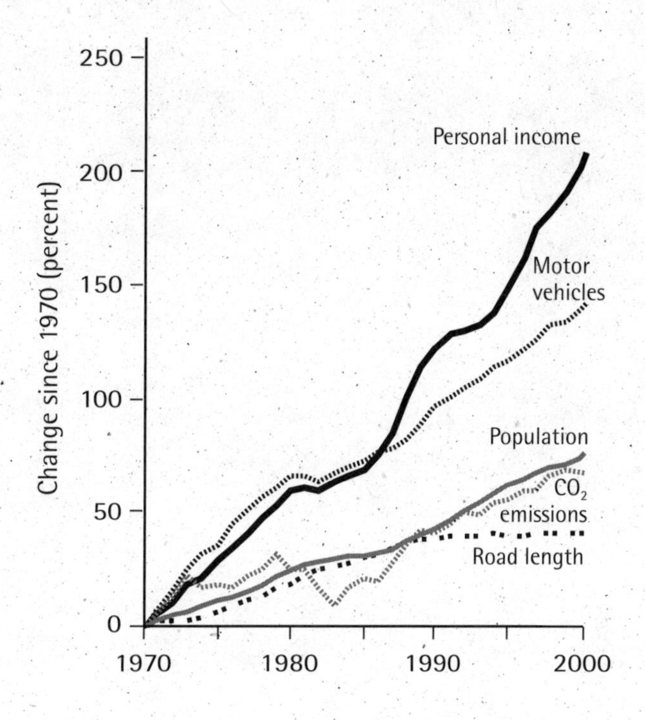

Figure 2. Since 1970, growth in the Northwest's income has far outpaced growth in environmental harm, such as road building and greenhouse gas emissions. Sources: see endnote 13.

HEALTH

indicator

By some measures, northwesterners are healthier than ever before. We live longer than our parents did, and our lifespans are increasing: a baby born in 2002 can be expected to live up to two months longer than a baby born in 2001 (see Figure 3). Yet troubling health trends remain. Many northwesterners lack health insurance, reducing their access to medical care. And deaths among young northwesterners, largely from traffic accidents and cancers, hold down the region's life expectancy substantially.[15]

As a measure of public health, life expectancy—the average number of years a newborn will live given current patterns of risk and mortality—is crude but effective. Effective because the healthier we are, the longer we tend to live; and crude because, in theory, life expectancy can increase even if people spend more of their lives with unsatisfactory health.

British Columbia boasts the longest life expectancy, and the largest gains in lifespan, of any part of the Northwest. A baby born in British Columbia in 2000 had an expected lifespan of 80 years and 4 months, about 2 years longer than a baby born a decade earlier. Life expectancy in the northwestern United States is rising more slowly. Between 1990 and 1999, average lifespans for newborns increased by 1 year and 5 months in Washington and by 6 months in Idaho. Overall, residents of the US Northwest could expect to live more than a year longer than the rest of Americans, and BC life expectancies exceeded the rest of Canada's by roughly a year.[16]

One factor that may extend the lifespans of British Columbians beyond their southern neighbors' is that Canadians have access to health

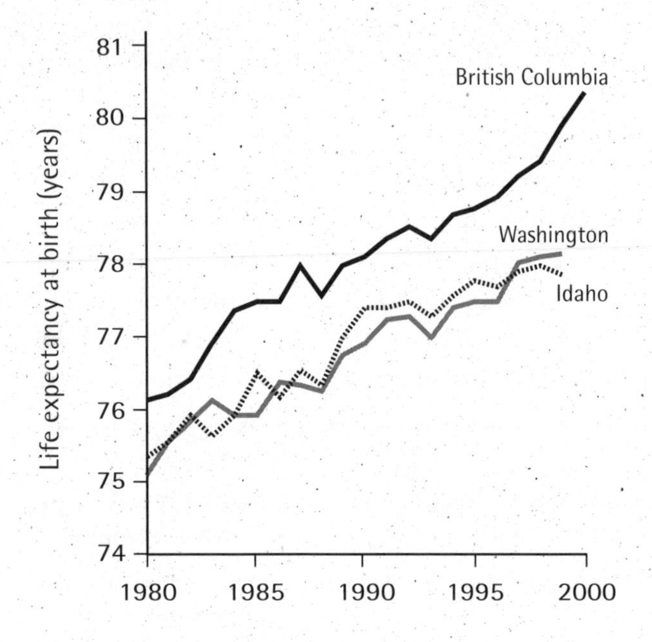

Figure 3. Northwesterners born in 2001 can expect to live longer than their older siblings, especially in British Columbia. Sources: see endnote 15.

care regardless of their income. South of the 49th parallel, as many as 1.4 million northwesterners go without health insurance. Canada's health system has given broader access to preventive care and narrowed disparities between rich and poor; wide income gaps are associated with poorer health.[17]

Deaths among the young disproportionately reduce life expectancy. Motor vehicles are the leading cause of death among people between the ages of 1 and 24 in the Northwest states, claiming 500 young lives each year, or nearly a third of all deaths in the age group. Cancer also claimed 120 young people each year in the northwestern United States from 1995 through 1998. In Washington State alone, doctors diagnose cancer each year in nearly 200 children under the age of 15.[18]

Public health agencies, academic institutions, and charities throughout North America track dozens, if not hundreds, of indicators of personal and public health. But they do not track them uniformly. Oregon's health department, for example, does not report life expectancy, even though calculating it from population and death records is fairly straightforward. This oversight is regrettable, because life expectancy may be the best single indicator of overall health.

INCOME

indicator

The last two decades were good to the pocketbooks of most northwesterners, but the economy disproportionately favored those at the top of the income ladder. Even as the high-tech boom of the 1990s minted legions of paper millionaires, the poorest northwesterners saw their incomes fall. By the time the boom wound down in 2000, many northwesterners were better off, but many others had never shared in the region's bounty.

Income inequality—a measure of how widely the fruits of prosperity are shared—has important implications for society. Where income inequality is high, societal ills tend to multiply: violence and property crimes increase, voting and public investments decrease, and health and life expectancy deteriorate. Social scientists believe that, as economic gaps widen, people increasingly segregate themselves. When disparities persist, shared values weaken, economic anxieties increase, and social cohesion frays.[19]

Despite the region's sustained economic boom, economic insecurity is high and growing. In the Northwest states, 180,000 more people lived below the poverty line in 2001 than in 1981. All told, more than 1.1 million people in Idaho, Oregon, and Washington—including more than 430,000 children—now live in poverty (see Figure 4). North of the border, British Columbia's wealth—a truer measure of economic well-being than income—is less evenly distributed than that of any other Canadian province. BC's millionaires, who make up only 3.3 percent of the population, own 35.5 percent of the wealth. Meanwhile, BC's poverty rate—which has more generous standards than its US counterpart—has climbed during the last 20 years, adding at least 170,000

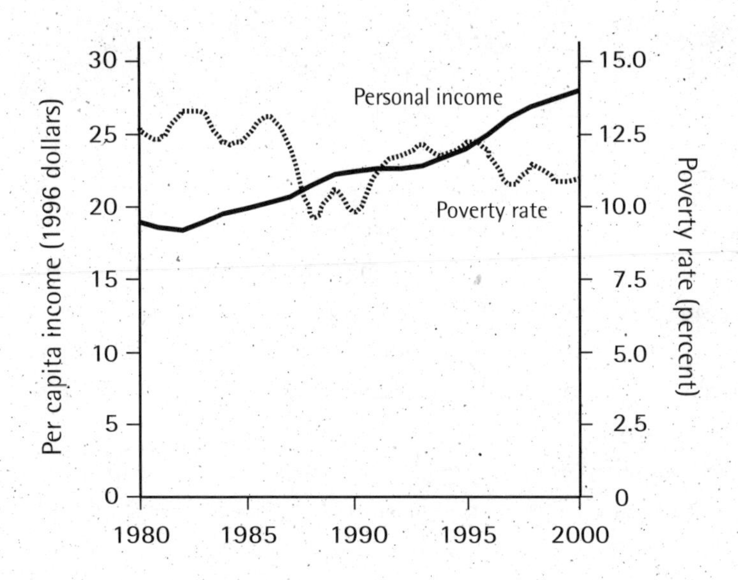

Figure 4. Despite surging income in the Northwest states, the poverty rate has declined little in the last 20 years. Sources: see endnote 20.

people to the ranks of the poor. British Columbia also counts 131,000 children among its poor.[20]

As average income grew over the past two decades, so did the disparities between the richest and poorest, and between the richest and the middle class. From the late 1970s to the late 1990s, the highest-earning fifth of US Northwest families saw their annual inflation-adjusted incomes grow, on average, by $38,000—enough for each well-off family to buy a new luxury sport utility vehicle every year. Over the same period, families in the middle saw an increase less than a tenth as large, while income for the poorest families actually declined (see Figure 5).[21]

If anything, income figures understate the region's economic disparities. Income measures exclude government benefits that go to the poorest members of society, such as food stamps, housing assistance, and school lunches. But they also exclude the far larger gains that accrue to the wealthy from investments, stock options, health care plans, and employer contributions to retirement funds. Income figures exclude taxes as well as tax breaks for housing, which overwhelmingly benefit the well-off. On balance, the poorest families may be marginally better off than income data alone suggest, while the richest families likely are dramatically richer.

In the Northwest, income disparities mean disparities in living standards. Food insecurity in the Northwest states is among the worst in the United States, even though it exists side-by-side with plenty: for every northwestern household earning $100,000 or more a year, another household has difficulty keeping enough food on the table. And housing trends, like hunger, exemplify the growing gap: there were more overcrowded dwellings in the Northwest states in 2000 than in 1980, while the number of second homes rose steadily over those same 20 years.[22]

The region's economic disparities did not grow uniformly over time or geography. Income gaps widened more quickly during the 1990s than during the 1980s, particularly in Oregon—the least equitable place in the Northwest—where in the 1990s the poorest families' incomes declined by 14 percent while the richest families' incomes soared by 38 percent. But in the past, British Columbia has dampened the impact of widening earning disparities through progressive taxes and government programs.[23]

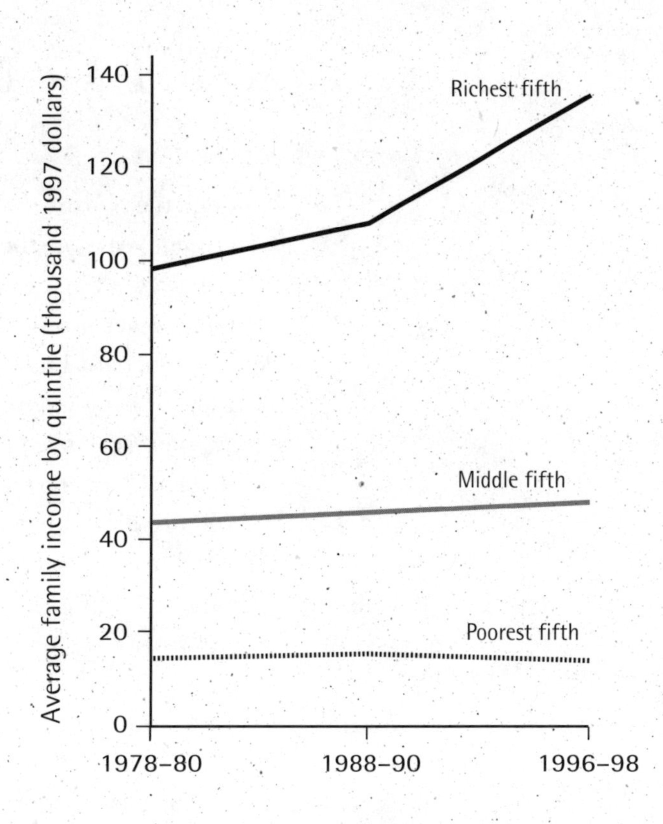

Figure 5. Twenty years of prosperity in the Northwest states mainly helped rich families. Sources: see endnote 21.

Broadly shared economic gains will be unlikely unless middle- and lower-income northwesterners can accumulate valuable assets, rather than just living month to month or hand to mouth. Wealth is perhaps the single most important determinant of financial security, since it cushions income shortfalls. Yet the changing distribution of wealth is among the most poorly measured of all financial trends in the region.

POPULATION

indicator

Although the Northwest's population grew more slowly in 2001 than in most recent years, it still increased by an estimated 197,000, enough to push the total above 16 million for the first time (see Figure 6).[24]

This slower growth is welcome partial relief from the strains that rising numbers of human inhabitants put on the Northwest. Expanding ranks of people worsen traffic and sprawl; increase energy use and solid waste; escalate emissions of greenhouse gases; and impose heavier tax burdens. New residents' demands on roads, schools, and other infrastructure drive up public expenditures more quickly than the new residents' tax payments fill public coffers. In Thurston County, Washington, for example, keeping up with growth saddles present residents with up to $1,200 in extra costs per year per taxpayer.[25]

Despite a booming economy in the late 1990s, population growth declined from its all-time peak of 43 new northwesterners per hour during 1990 to 23 per hour during 2000 and an estimated 22 per hour during 2001 (see Figure 7). The Northwest's growth rate since 1990 has been twice the North American rate, faster than India's, and almost equal to Egypt's. If this rate continues, northwesterners will double to 32 million by 2040; Idaho, the fastest-growing part of the region, will double by 2032 (see box, page 18). But if the last two years signal an enduring population slowdown, those milestones will recede. If growth stays at 2000–01 rates, for example, population doubling would delay by two decades—a much needed respite but still an ominous trajectory for the region's future.[26]

Gradual declines in the birthrate, especially among teens, helped usher in the more measured pace of recent growth, but the main cause

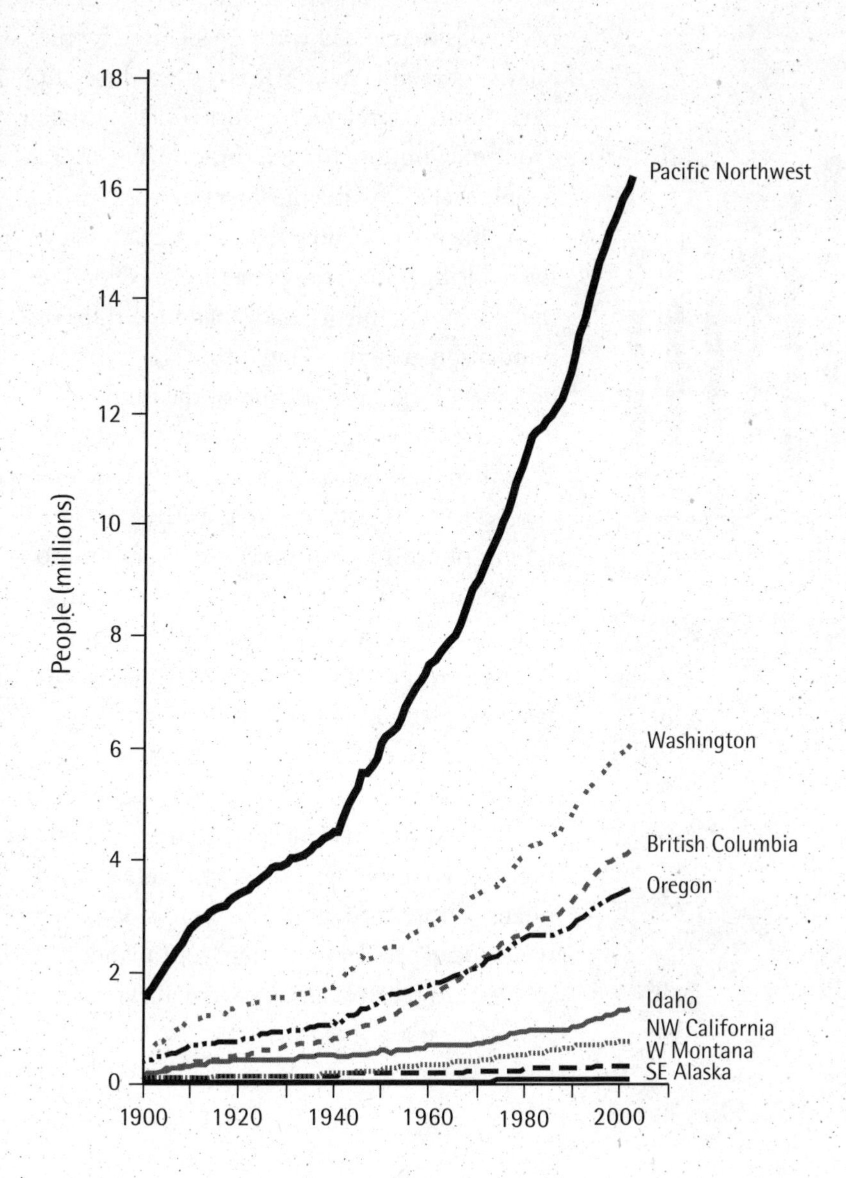

Figure 6 (this page). The Northwest's population has doubled since 1965. Sources: see endnote 24.

Figure 7 (next page, top). The Northwest's population grew by an estimated 22 people per hour in 2001. Sources: see endnote 27.

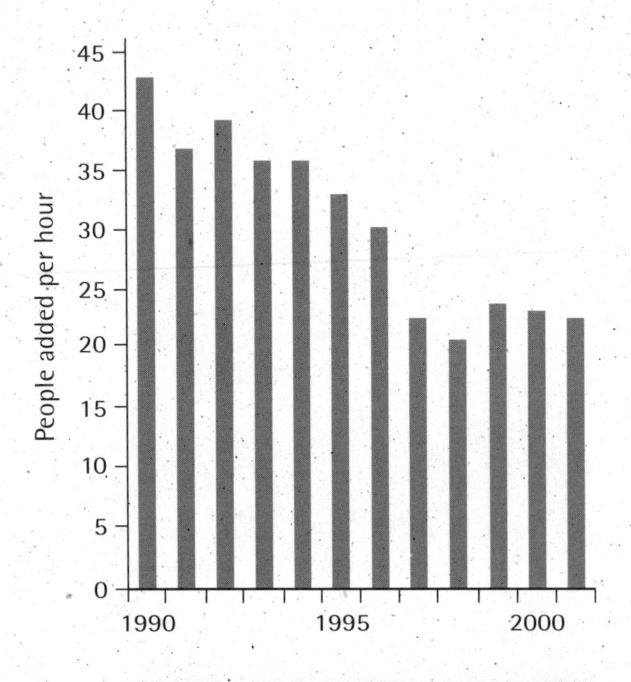

When We'll Double

Idaho	2032
British Columbia	2039
Washington	2040
Pacific Northwest	2040
Oregon	2041
Western Montana	2041
Northwestern California	2057
Southeast Alaska	2110

Sources: see endnote 24.

was slower migration into the region. Migration tends to come in waves, generated by the relative strength of job markets in different parts of the continent. California's late-1990s economic boom, for example, drew more people there, slowing the flow into the Northwest. Births, in contrast, take place steadily. Even during the 1990s, when migrants flooded the Northwest, births accounted for a third of population growth; as migration slows, births become a larger proportion of the total.[27]

Root causes of the region's fast population growth include not only individuals' conscious choices but also public policy. The region subsidizes migration, for example, by failing to recover from developers the full costs of infrastructure. And a large share of the region's births—37 percent in Washington—result from accidental pregnancies.[28]

In fact, were the data available, the percentage of births from unintended pregnancies might make a better indicator of sustainability than total population. If all births were wanted births, population growth would slow, even while social problems ranging from abandoned children to teen pregnancy were alleviated.[29]

SMART GROWTH

indicator

By some measures, "smart growth" is catching on. Roughly 32 percent of residents in the Northwest's three great metropolitan areas lived in compact, smart-growth communities at the time of the last census, up from 27 percent a decade earlier. These compact neighborhoods accounted for slightly more than half of the three cities' population increase over ten years. Low-density sprawl, where residents depend on cars for virtually all transportation, accounted for the rest of metropolitan growth.[30]

This news conceals three distinct growth trends. Seattle sprawled: low-density residential areas made up three-fifths of Seattle's growth during the 1990s. Vancouver, BC, grew denser: high-density neighborhoods contained 80 percent of the city's growth from 1986 through 1996. And Portland grew in a fairly even mix between car-dependent sprawl and transit- and pedestrian-friendly densities (see Figure 8).

Density—population per acre or hectare—is the main determinant of how much people depend on their automobiles (see box, page 21). At low densities, residents need cars to get to work, stores, and basic services; anyone without a car is stranded. Higher densities give residents more transportation choices. Scholars of the urban form have found that, in areas with more than 12 people per acre (30 people per hectare), public transit becomes cost-effective, bus ridership increases, vehicle ownership dips, and total gasoline consumption falls. Above roughly 40 people per acre (100 people per hectare)—at typical downtown densities—destinations are close enough that walking and biking flourish, driving decreases substantially, and as many as a third of households do not own a car at all.[31]

dyadic sexual regulation scale earned run average emotional intelligence quotient

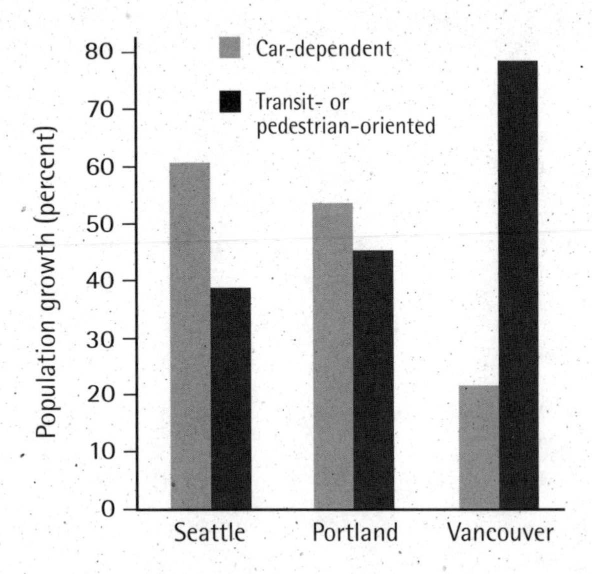

Figure 8. Seattle has sprawled while Vancouver has grown more compact. Sources: see endnote 30.

Sprawl brings drawbacks besides car dependence. Sparsely populated neighborhoods require more pavement per resident and have higher costs for fire, ambulance, police, and trash services as well as for infrastructure such as electricity, phone lines, and television cables. Sprawl isolates residents, reducing contact between neighbors and fraying social bonds. It may even worsen health, by promoting sedentary lifestyles, smog, and traffic accidents (see "Health").[32]

SEATTLE

At the start of the 1990s, the Seattle-Tacoma metropolitan area—stretching from Everett in the north through Seattle, Redmond, and Bellevue and south to Tacoma and its neighboring suburbs—was home to 2.7 million residents. By 2000, its population had grown to 3.2 million. Seattle accommodated most of this growth by sprawling: the number of people living at car-dependent densities increased by about 320,000 over the decade—more than 60 percent of the region's population growth. By the end of 2000, 2.5 million Seattle-area residents—three-quarters of the total—lived at car-dependent densities.

Seattle's sprawling suburbs consume a disproportionate share of the region's landscape (see Figure 9, page 27). In 2000, the Puget Sound region's car-dependent neighborhoods occupied nearly five times as much land per person as its transit- and pedestrian-friendly communities. Between 1990 and 2000, the area occupied by Seattle's car-dependent suburbs increased by 75,000 acres (30,000 hectares)—an area larger by a third than the city of Seattle itself.

Still, the news was not all bad. Pedestrian-friendly neighborhoods attracted many residents; the number of people living in communities such as Belltown and First Hill grew more than 80 percent, thanks largely to construction of new apartments and condominiums. Overall, the number of people living in transit- and pedestrian-friendly neighborhoods grew 35 percent over the decade, more than twice the growth rate of car-dependent neighborhoods.

Are You Dense?

Car-dependent	< 12 people per acre (< 30 people per hectare)
Transit-oriented	12–40 people per acre (30–100 people per hectare)
Pedestrian-oriented	> 40 people per acre (>100 people per hectare)

Sources: see endnote 30.

PORTLAND

The Portland metropolitan region—from Vancouver, Washington, in the north through Portland and its surrounding towns and as far south as Salem—sprawled much less than the Seattle area (see Figure 10, page 28). The region added about 470,000 people during the 1990s, and compact neighborhoods accounted for nearly half that increase. The number of Portland-area residents living at transit- and pedestrian-friendly densities increased from about 310,000 in 1990 to 530,000 in 2000, a high-density growth rate nearly double Seattle's. At the same time, the number of people living in the least dense suburbs (less than 5 people per acre, or 12 people per hectare) actually fell.

Oregon's growth management laws apparently limited sprawl and encouraged denser neighborhoods. They also helped maintain clear boundaries between residential and rural lands; within Oregon, land with little or no suburban development separated population centers. In Portland's Washington State suburbs, in contrast, sprawl predominated.

VANCOUVER

In 1986, greater Vancouver, BC—comprising Vancouver, Richmond, New Westminster, and surrounding towns—had by far the highest densities among the Northwest's three large cities. By 1996 it had grown even denser (see Figure 11, page 29). The region added 450,000 new residents over the decade, and growth at transit- and pedestrian-oriented densities accounted for 80 percent of this increase. As a result, the share of people living in dense neighborhoods increased from a slim minority (49 percent) to a clear majority (57 percent). At last count, nearly half the area's residents lived at transit-friendly densities, and one in ten lived in a pedestrian-centered neighborhood.

The number of Vancouver residents living at car-dependent densities increased by about 100,000 over the decade—a sprawl rate about a third as fast as Seattle's and about two-fifths as fast as Portland's. Still, this increase in car dependence strained the region's roads and community infrastructure.

Vancouver and, to a lesser extent, Portland show that low-density suburbs are not an inevitable consequence of metropolitan growth. The deciding factor seems to be public policy: strong growth management laws help rein in sprawl. Population density, therefore, reveals both how efficiently we use land to meet our needs, and how effectively we use politics to serve our interests.

PAVEMENT

indicator

New development consumed nearly an acre of land every hour in the Northwest's three largest cities during the 1990s, expanding the footprint of the built environment by 12 square miles (32 square kilometers), on average, every year.[33]

Impervious surface—rooftops, sidewalks, roads, driveways, patios, and parking lots—is a proxy for both the physical scale and the ecological impact of northwesterners' real estate development. The benefits of buildings and pavement are clear enough, but their costs are mostly hidden. Pavement turns habitat for living things into habitat for cars. It can also turn rainstorms into chemical spills: water running off streets and parking lots carries traces of oil, pesticides, and other toxins. Likewise, making more of a watershed impervious to water worsens floods, which erode soils, clogging streams with sediment. Imperviousness also slows the recharge of aquifers, lowering water tables and raising stream temperatures. All these changes diminish water supply, harm water quality, and undermine aquatic ecosystems.[34]

At densities of one house per acre, impervious surface covers from 10 to 15 percent of the landscape, and streams begin to deteriorate; the Northwest's sensitive coho salmon rarely inhabit watersheds where impervious surface exceeds this level. Where impervious surface covers 25 percent of a watershed, streams can become inhospitable to aquatic life. Because further development adds relatively little ecosystem damage, human growth that concentrates new pavement and buildings in already urbanized areas is best for nature.[35]

No institution—public, private, or academic—directly monitors the amount of land consumed by buildings and pavement in the Northwest.

Analysis of satellite imagery, however, provides some estimates: images from a single year give a snapshot of development's scale, while comparisons across different years show its spread. Although these estimates do carry some uncertainty, comparison of satellite images taken a decade apart yields a bounty of information on where development is taking place and how much of the landscape it is devouring.

All three of the Northwest's big cities grew by roughly the same number of people during the 1990s. And in all three, the amount of land fully (upward of 80 percent) or partially (roughly 15–80 percent) covered by impervious surface increased by about a tenth. But in Seattle, which started with a much bigger footprint, the amount of land affected by new pavement and buildings was nearly double that of Vancouver and significantly larger than in Portland.[36]

SEATTLE

In greater Seattle, both the most populous and most expansive of the Northwest's cities, partially impervious land area grew, between 1988 and 1999, by 53 square miles (138 square kilometers) at the expense of previously undeveloped land. An additional 7 square miles (18 square kilometers) became fully impervious. In all, the region lost 10 acres (3.9 hectares) to development every day over the period.

Much of Seattle's development occurred on the metropolitan outskirts: south around Tacoma; east in Issaquah and Maple Valley; and north in Mill Creek (see Figure 12, page 30). Scattered development was common, particularly on rural Kitsap Peninsula and east of Lake Washington. And this sprawling growth meant that development consumed more land per capita in Seattle than in either Portland or Vancouver.

Washington's 1990 growth management law required cities and towns around Puget Sound to develop urban boundaries and growth management plans. Most of Seattle's new impervious surface over the decade was built within the region's growth boundaries. Despite this apparent success, however, Washington's growth management plans may not have channeled sprawl but simply anticipated where it would occur. Future analyses may better determine how effectively the law prevents sprawl.

PORTLAND

Portland's development was more contained and compact than Seattle's (see Figure 13, page 31). From 1989 through 1999, development encroached on nearly 46 square miles (120 square kilometers) of land, or roughly 8 acres (3 hectares) every day.

The pattern of Portland's impervious surface differs markedly from Seattle's. Where Seattle-Tacoma shows patchy, spread-out development, Portland is characterized by well-bounded cities and suburbs separated by undeveloped land. Much of Portland's new impervious surface resulted from infill—redevelopment within existing urban and suburban areas—rather than uncontained sprawl. As a result, despite Portland's considerable population growth, new suburbs did not finger their way into Oregon's rural land.

Suburban sprawl does not—as Vancouver, BC, shows—inevitably follow from population growth

Oregon's strong growth management laws almost certainly played a role in keeping Portland's sprawl in check. In Portland's suburbs north of the Columbia River, the patchy development pattern more closely resembles that of Puget Sound than the rest of greater Portland.

VANCOUVER

From 1987 through 1999, Vancouver's footprint expanded by about 26 square miles (67 square kilometers), or roughly 4 acres (1.5 hectares) every day. Suburbs or other medium-intensity development covered roughly 24 square miles (61 square kilometers) of new land, while impervious surface fully covered a little over 2 square miles (5 square kilometers).

Vancouver's new construction, like Portland's, concentrated in or near areas of existing development. Infill was common, as was new development adjacent to previously developed residential zones. As in Portland, boundaries between town and farmland are crisp, and undeveloped land near the city has been preserved (see Figure 14, page 32).

Though the Vancouver-area population grew by a third between 1986 and 1996, the land devoted to cities and suburbs increased by at most 11 percent over the 12 years covered by satellite images. By confining new construction to infill and areas adjacent to existing development, Vancouver preserved open space while encouraging dense, compact communities.

Pavement's extent does not fully measure its impacts. A more refined indicator would guage efficiency: how much impervious surface exists per person, how much is added for each new resident, and whether new pavement fragments habitat or degrades watersheds. Still, the growing footprint of development gives a fair approximation of its effects, and future refinements in satellite imaging and analysis will likely yield better estimates of pavement's impact on the landscape.

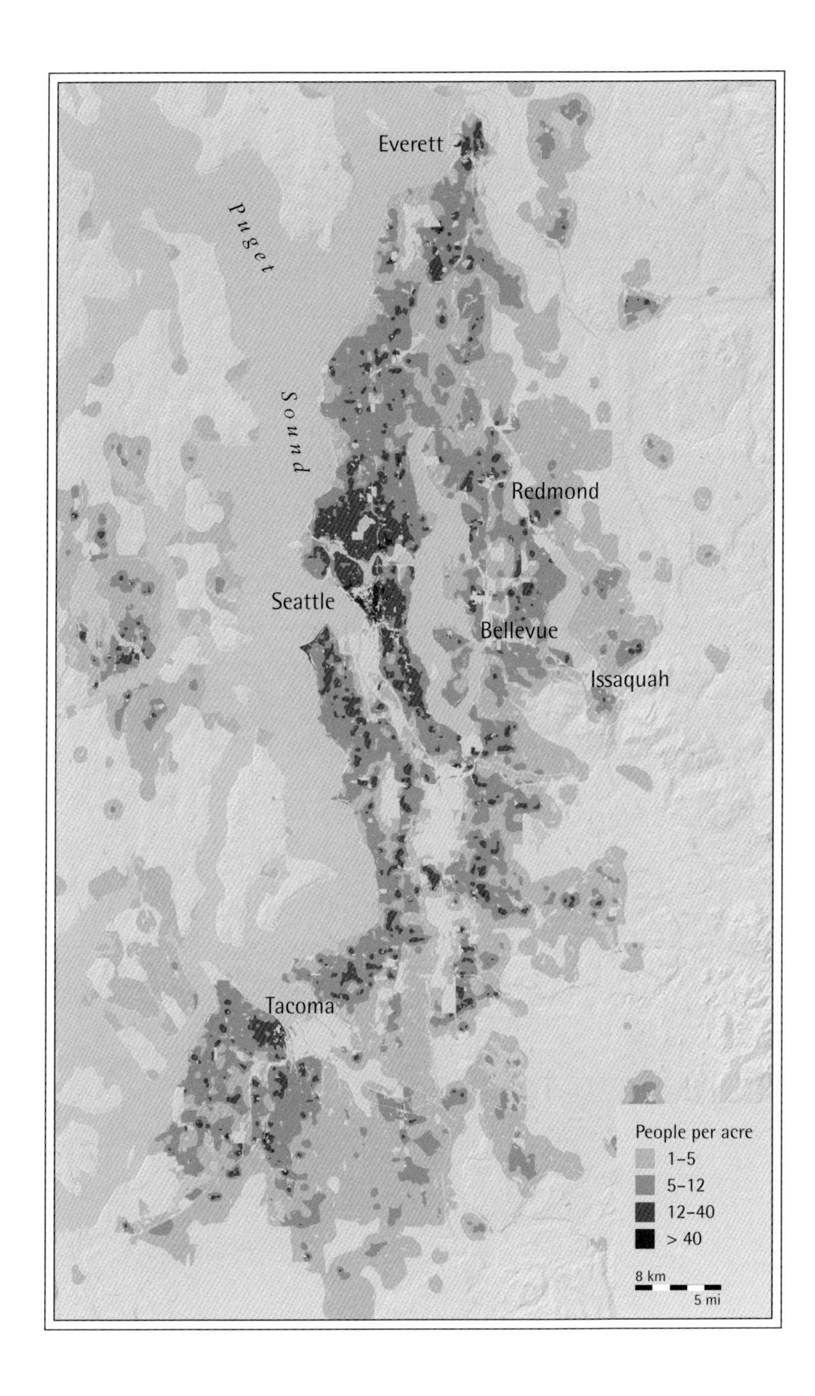

Figure 9. The Puget Sound region's population growth has taken the form of low-density sprawl.
Map by CommEn Space, Seattle; see endnote 30.

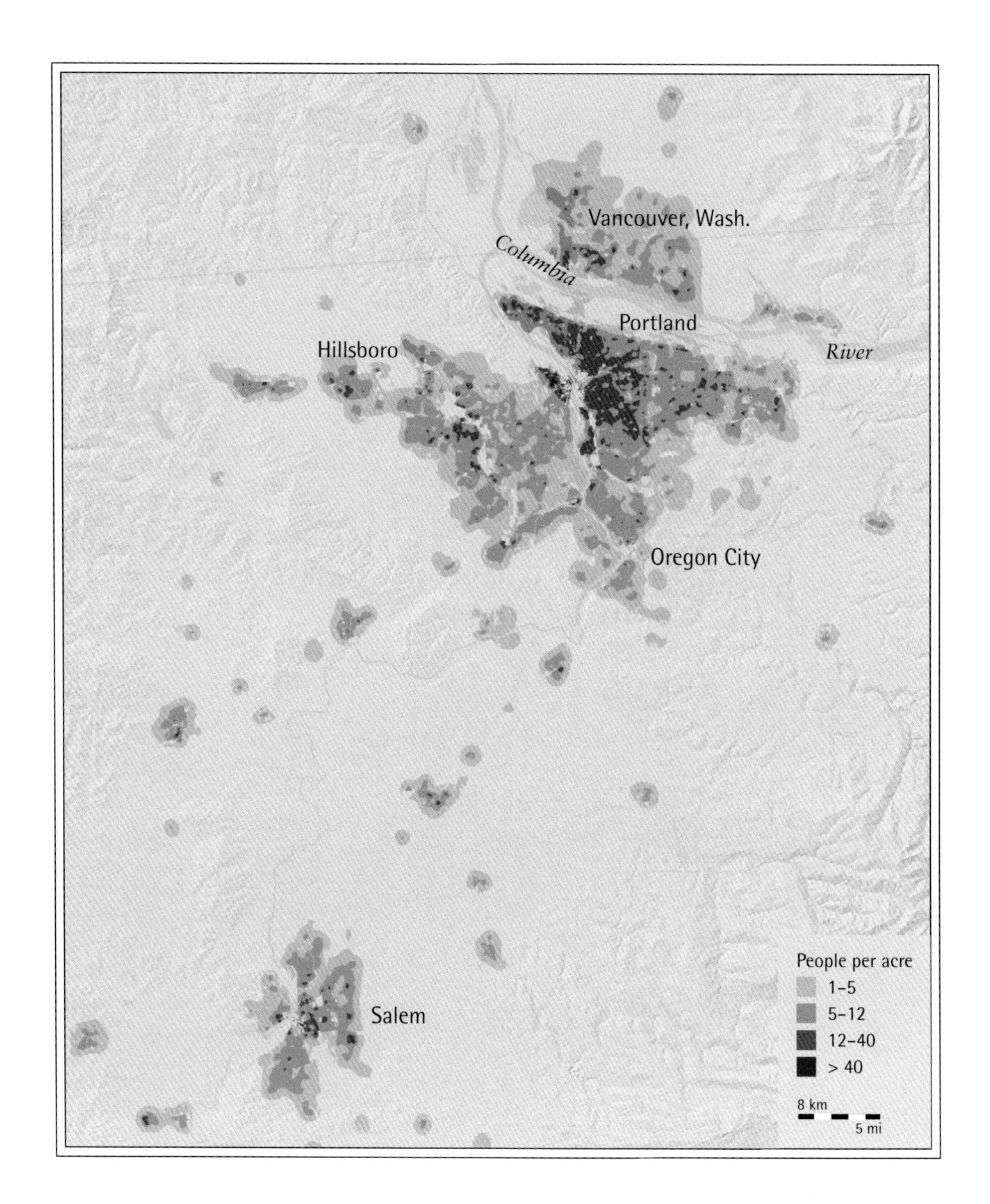

Figure 10. Compact communities accounted for nearly half of Portland's population growth. Map by CommEn Space, Seattle; see endnote 30.

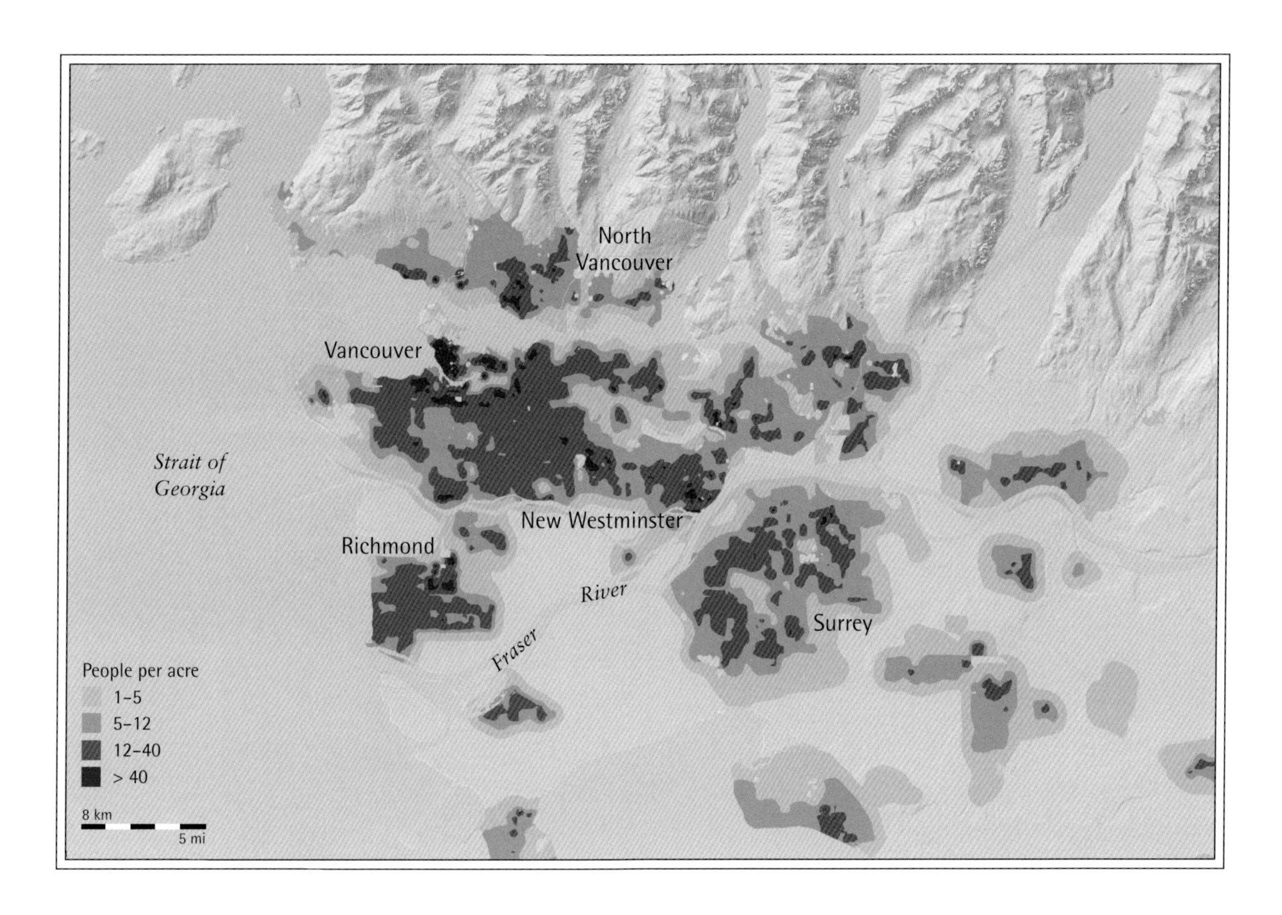

Figure 11. Greater Vancouver, BC, is by far the region's densest major city.
Map by CommEn Space, Seattle; see endnote 30.

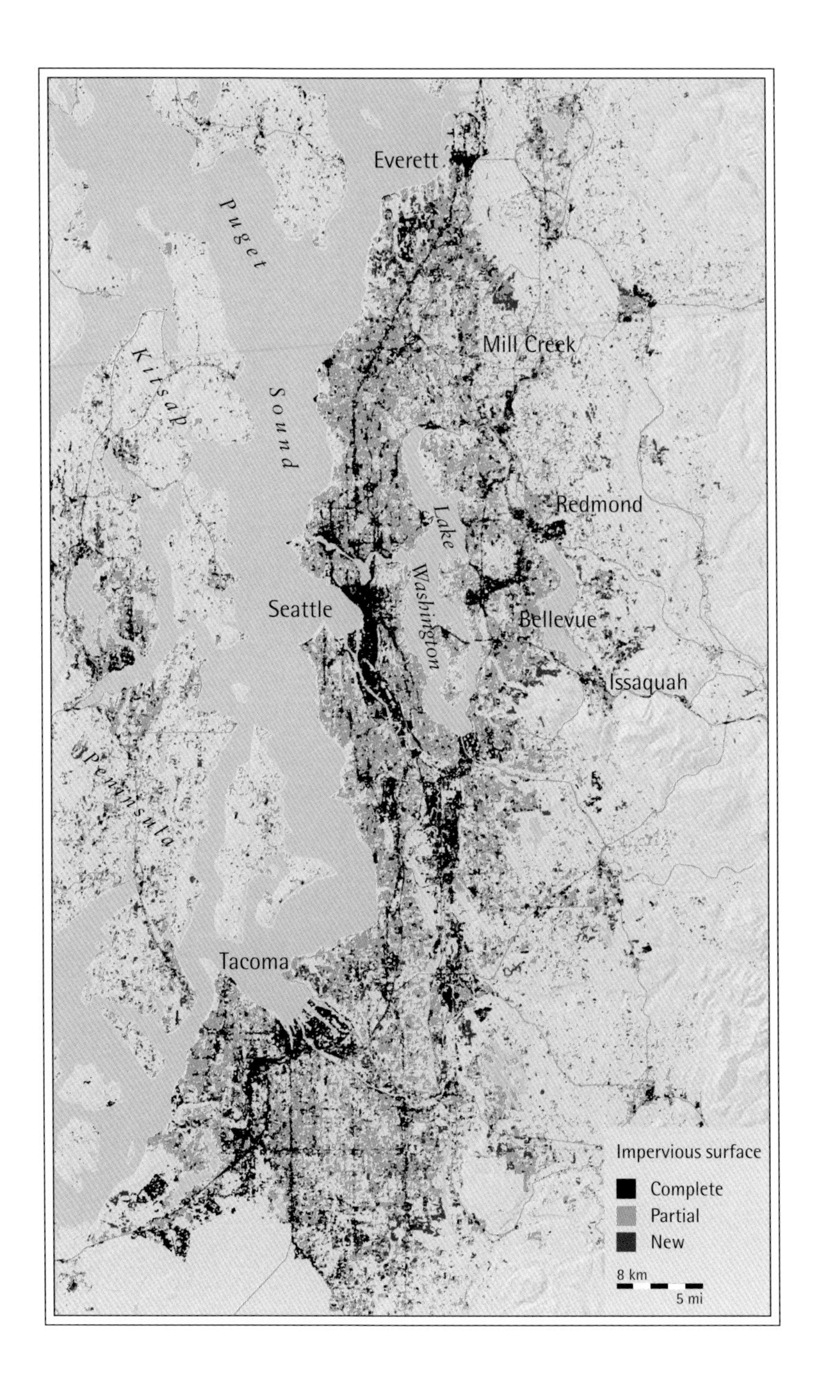

Figure 12. Seattle lost ten acres to development every day from 1988 through 1999. Map by CommEn Space, Seattle; see endnote 33.

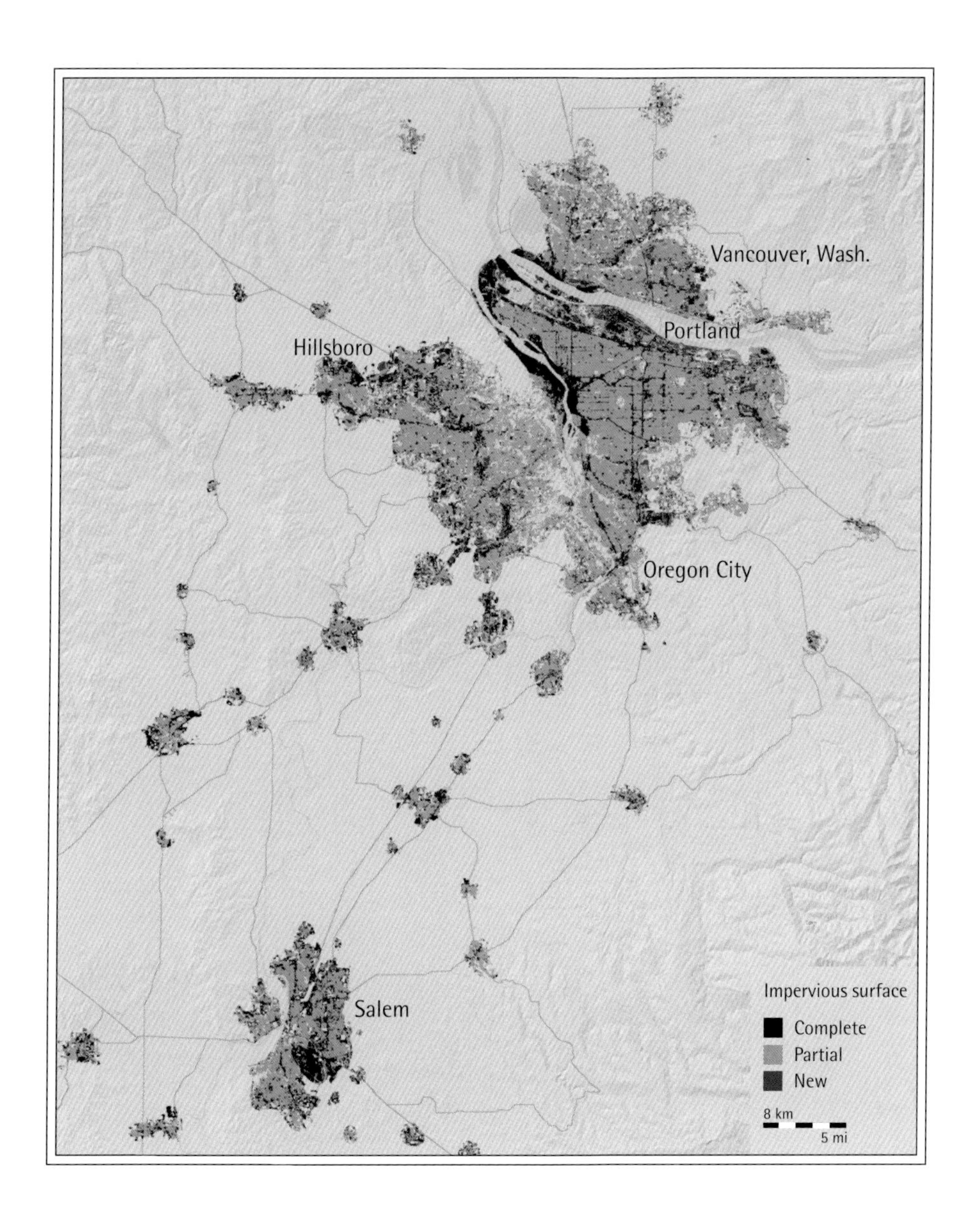

Figure 13. From 1989 to 1999, new pavement in Portland came from redevelopment within existing urban areas, yielding a more compact footprint than Seattle's. Map by CommEn Space, Seattle; see endnote 33.

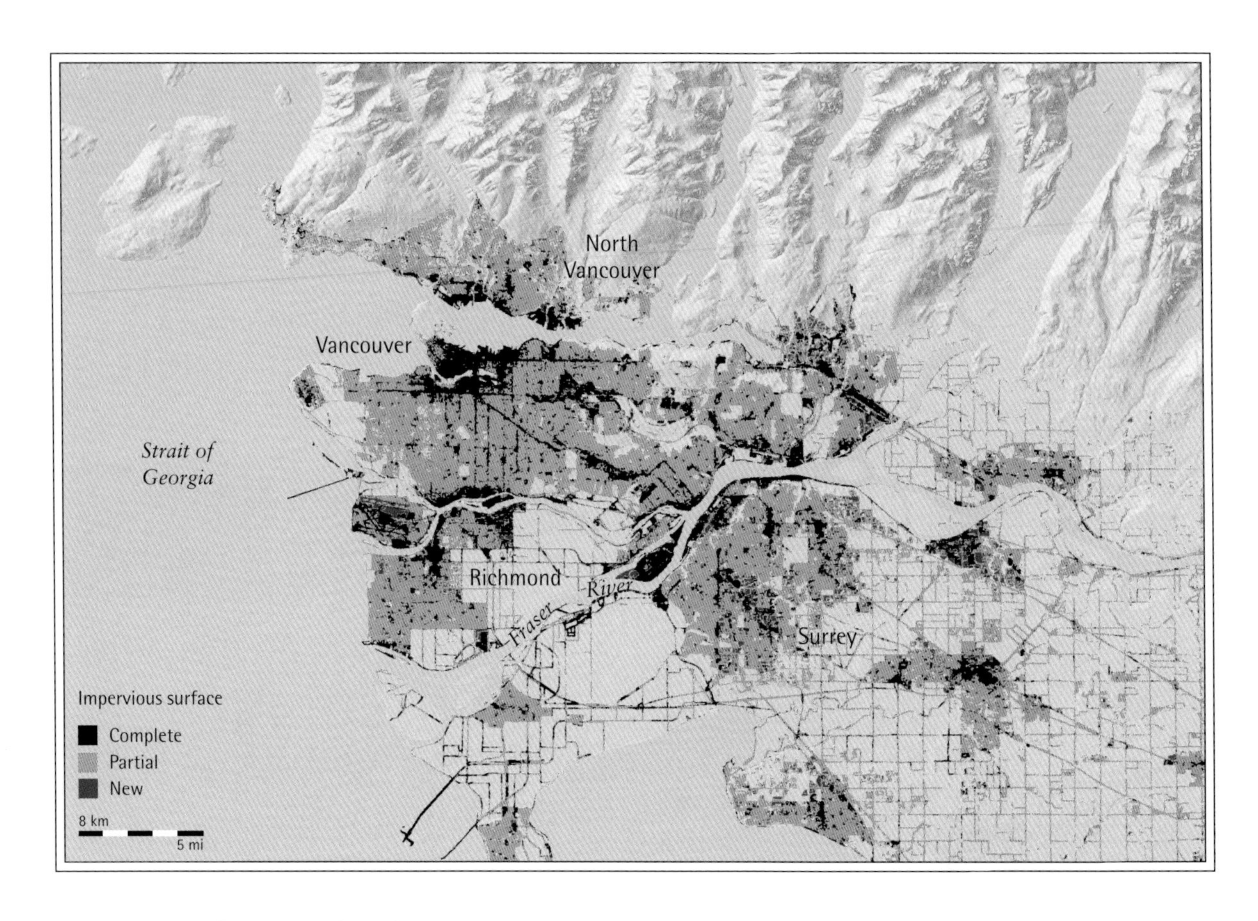

Figure 14. From 1987 through 1999, Vancouver's development preserved open space while encouraging dense, compact communities.
Map by CommEn Space, Seattle; see endnote 33.

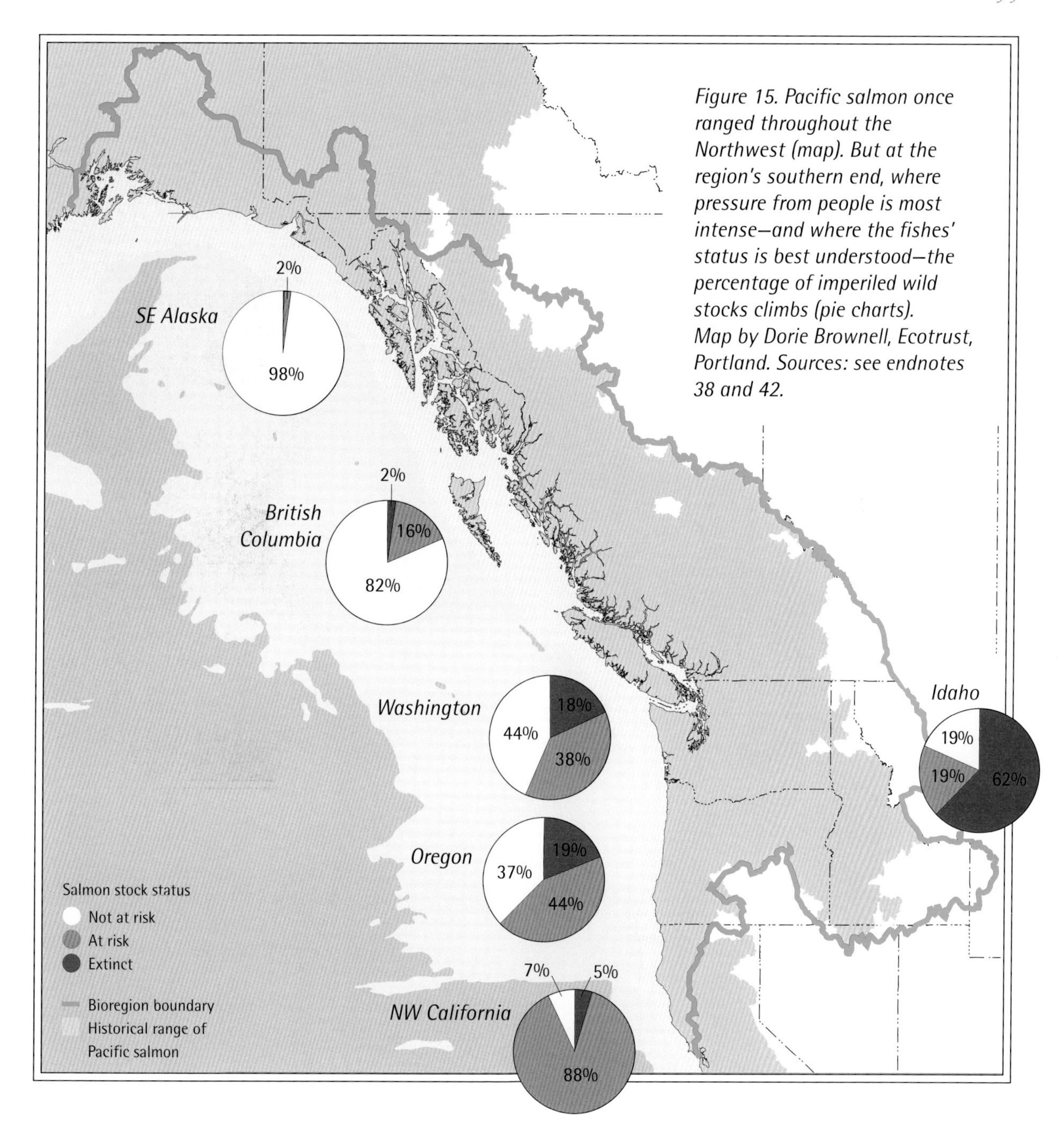

Figure 15. Pacific salmon once ranged throughout the Northwest (map). But at the region's southern end, where pressure from people is most intense—and where the fishes' status is best understood—the percentage of imperiled wild stocks climbs (pie charts). Map by Dorie Brownell, Ecotrust, Portland. Sources: see endnotes 38 and 42.

SALMON

indicator

Like many years, 2001 was a time of both boom and bust for Pacific salmon. In Washington and Oregon, adult fish returning to fresh water to spawn were more abundant than they had been in decades. Coastal salmon fisheries opened for the first time since 1977, and by year's end, more than 2 million fish, most of them hatchery raised, had scaled the ladders of the Columbia River's Bonneville Dam. Climate played a role in the boom: a recent shift in northern Pacific atmospheric and oceanic currents benefited the region's southern runs, though at the expense of Alaskan fisheries. For juvenile salmon, in contrast, the climate made for an abysmal year. Drought caused record-low river flows, and dam managers used most of the available water to produce electricity, reserving little for salmon. Of those Columbia and Snake River juveniles that managed to reach the ocean, more than half made the journey in barges. Hundreds of thousands of young fish remaining in the rivers perished, and the typical downstream migration time of many others doubled.[37]

These ups and downs may dominate headlines, but for salmon, they are commonplace. Salmon populations fluctuate naturally from decade to decade. Longer-term trends matter more, and these trends are plain: over the past 150 years, the range, diversity, and abundance of wild salmon have declined dramatically.[38]

The status of wild salmon stocks is a telling indicator of ecological health. Besides humans, no creature penetrates the Northwest so completely. In a lifetime, a single fish may be a denizen of mountain, forest, desert, farm, city, estuary, and ocean; its migration unites the lands and waters through which it travels. Salmon reflect the cumulative stresses on the waters that support them—and logging, ranching, farming, mining, cities, dams, pollution, global warming, and overfishing have degraded these waters. Wild salmon also face serious threats from hatchery- and farm-raised fish. Rather than replenishing wild runs, hatchery fish weaken and compete with wild stocks, even as abundant hatchery fish promote overfishing.[39]

Wild salmon both reflect and contribute to the health of their watersheds. When adult fish return to their natal streams to spawn and die, they bring with them a wealth of ocean-derived nutrients. Their carcasses feed more than 135 different species of animals, as well as the tiny organisms at the base of aquatic food chains—which in turn feed juvenile salmon. Before the 1930s, grizzly bears derived between 35 and 91 percent of their bodies' carbon and nitrogen from the sea, borne inland by salmon; BC researchers estimate that 70 percent of black bears' annual protein still comes from salmon. And salmon fertilize the woods: streamside trees and shrubs in Alaska gain roughly 25 percent of the nitrogen in their leaves from offshore, and Sitka spruce grow more than three times as fast along spawning streams as along streams without salmon. South of British Columbia, salmon-borne nutrients have plummeted by 90 percent or more.[40]

Forest plants and animals derive much of their nutrition from the sea, borne inland by salmon

Despite the comparatively strong adult runs of 2001, salmon abundance has fallen far below historical levels. For every 50 salmon the Columbia River basin supported a century and a half ago, in recent years it has supported perhaps 7, only 1 or 2 of which are wild. Even in the undammed Fraser River of British Columbia—where hatcheries are absent and salmon still swim in what one writer calls "jaw-dropping numbers"—recent runs come to roughly a third of their former size.

Such declines have led to shrinking harvests, lost fishery income, and dislocated communities.[41]

Like their abundance, the salmon's geographic range has shrunk. The seven species of Pacific salmon—chinook, chum, coho, pink, sockeye, steelhead, and sea-run cutthroat trout—once occupied nearly the entire Northwest; they have now disappeared from 40 percent of their former range in Washington, Oregon, Idaho, and California. Salmon fare worst in the southern part of the region, where pressure from people is most intense, but also where the fishes' status is best known (see Figure 15, page 33). Only about half of Alaska's and British Columbia's salmon stocks (excluding sea-run cutthroat) have been evaluated, and of these, 2 percent in southeast Alaska and 18 percent in British Columbia are depressed or extinct. In Washington, Oregon, Idaho, and northern California, the share of extinct and imperiled stocks climbs to well over half. Across the Northwest, at least 214 salmon runs are now extinct, and more than 1,000 are likely at risk.[42]

The decline of salmon signals failing health and productivity in a wide range of ecosystems

Shrinking range and threatened stocks are not the only, or perhaps even the most important, measures of deteriorating Northwest lands and waters. Historically, a mosaic of survival strategies enabled different groups of salmon to use different river and stream habitats at different stages in their life cycles. These strategies, known as life histories, lent resilience: if one food source or habitat was damaged, the entire run did not perish. Today, as river and stream habitats fray, many life histories are disappearing. In the Columbia River basin, a vast array of survival strategies has been winnowed to a handful, and only one run still spawns naturally in the main river. As life histories dwindle, so do salmon's survival options in a changing environment. Biologists are just beginning to catalog salmon life histories in Northwest rivers, but the patterns are disappearing faster than scientists can count them.[43]

The decline of salmon signals failing health and productivity in a wide range of ecosystems. The important thing is what we make of this signal. We could act on it by carrying out a broad recovery in both aquatic systems and the landscapes that support them—a difficult task but one that will bring lasting rewards. Instead we have responded to declining numbers of fish by manufacturing more of them in artificial hatcheries. This response is flawed: if the canary in the coal mine is dying, breeding more canaries does nothing to purify the air.[44]

CARS & TRUCKS

indicator

Growth of the Northwest's motor vehicle fleet slowed in 2001, adding an estimated 110,000 cars and trucks to the region's roads—half the average annual increment of the past decade. Still, the increase brought the region's tally of cars and trucks to more than 12 million (see Figure 16). With so many vehicles, the Northwest has enough seat belts to take itself and everyone in the state of California for a drive. Clearly, scarcity is not a problem; congestion is.[45]

Preliminary figures for 2001 show that the number of vehicles per northwesterner has remained fairly flat since 1990, after rising in earlier years. In 1963 the Pacific Northwest had one vehicle for every two people: everyone could get in a car or truck and no one would have to sit in the backseat. By 1990 the region had 83 vehicles for every 100 people, and—with compact communities growing faster than sprawling ones—increases in car and truck numbers slowed to roughly the same pace as population growth. (Vehicles outnumbered licensed drivers by the 1970s; today, every driver could hit the road and more than a million vehicles would still be parked.)[46]

Idaho, with its large rural population, has the most motor vehicles per person in the region: 95 per 100 residents. In Oregon, the figure is 89; in Washington, 85; and in British Columbia, with less purchasing power and more pedestrian- and transit-centered cities, the figure is just 70.[47]

That the region's vehicular numbers have stopped rising with income—despite galloping economic growth during the 1990s—is a welcome development, because cars and trucks, though exceptionally useful, are among the Northwest's principal environmental offenders.

They cause more air pollution and greenhouse gas emissions (see "Greenhouse Gases") than any other single source. They bind the regional economy to distant oil wells; strand urban northwesterners in some of the worst traffic in North America; take the lives of 2,000 northwesterners each year, on average, in collisions; consume as much as half of metropolitan land area for roads and parking; and facilitate sprawling suburban growth (see "Pavement"). The number of vehicles, therefore, is a rough indicator of the long-term impacts of the Northwest's transportation system on the region's livability.[48]

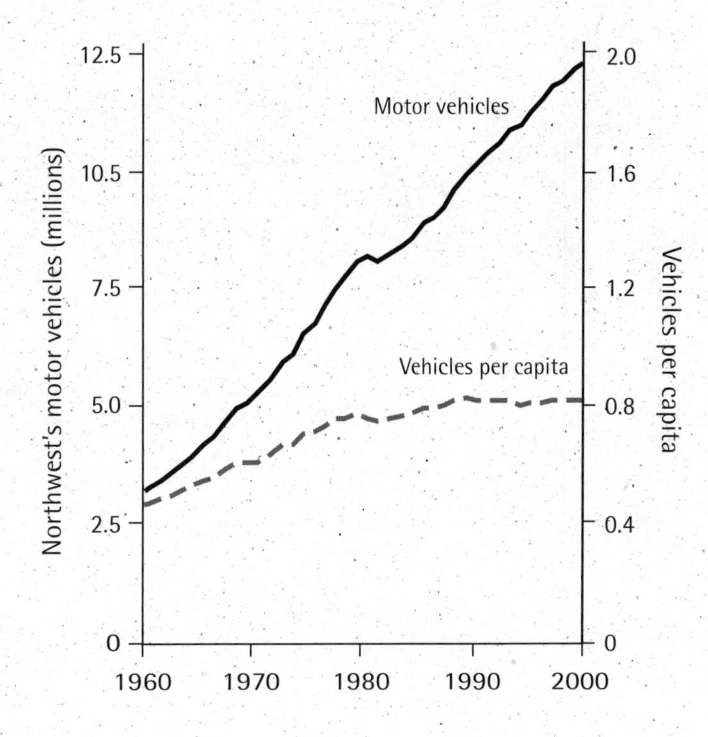

Figure 16. Cars and trucks continue to proliferate in the Pacific Northwest, but vehicles per capita stabilized in the 1990s. Sources: see endnote 45.

If the stabilization of vehicles per capita is good news, the size of those vehicles is bad news. Fuel-guzzling trucks—including heavy trucks and light trucks such as pickups, minivans, and sport utility vehicles—are rapidly outnumbering passenger cars (see Figure 17). Idaho already has more trucks than cars; Oregon will likely cross that threshold in 2002, and Washington is close behind. In British Columbia, the passenger vehicle fleet has been putting on weight rapidly—and upping its fuel appetite—as trucks proliferate and lighter cars dwindle (see Figure 18).[49]

The number of motor vehicles is an imperfect indicator of their environmental impact. Cars are readily countable—indeed, by law their

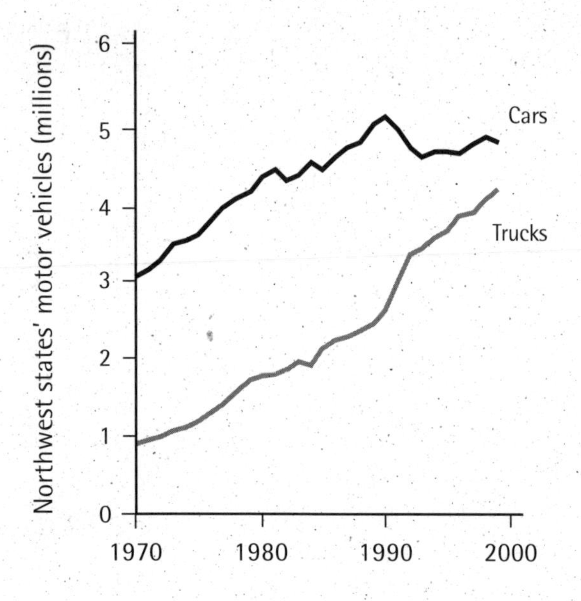

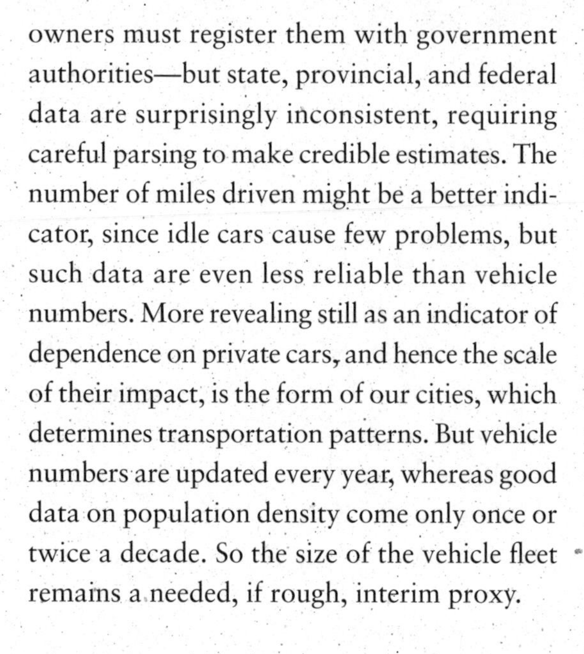

owners must register them with government authorities—but state, provincial, and federal data are surprisingly inconsistent, requiring careful parsing to make credible estimates. The number of miles driven might be a better indicator, since idle cars cause few problems, but such data are even less reliable than vehicle numbers. More revealing still as an indicator of dependence on private cars, and hence the scale of their impact, is the form of our cities, which determines transportation patterns. But vehicle numbers are updated every year, whereas good data on population density come only once or twice a decade. So the size of the vehicle fleet remains a needed, if rough, interim proxy.

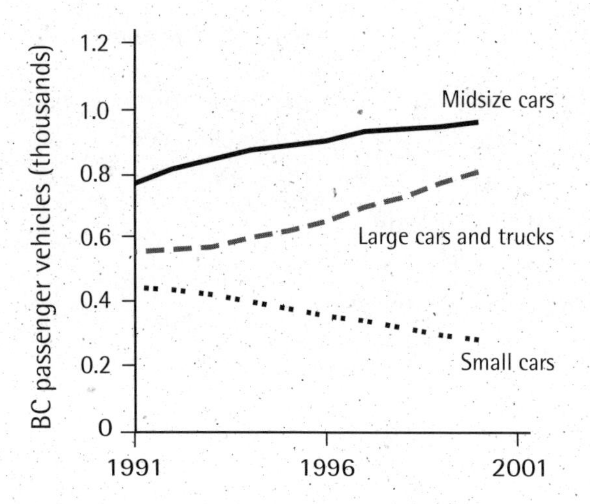

Figure 17 (top). If present trends continue in the Northwest states, trucks–including SUVs and minivans–will soon outnumber cars.
Sources: see endnote 45.

Figure 18 (bottom). In British Columbia, passenger vehicles are getting much bigger.
Sources: see endnote 45.

kinsey scale of sexual orientation life roles inventory-value scales life satisfaction scale

ROADS

indicator

After three decades of steady growth, the Northwest's streets, highways, and US national forest roads almost stopped proliferating in the last decade (see Figure 19). Northwesterners continued building streets and highways as in previous years, but outside of British Columbia, they reduced the pace of construction tenfold in national forests and obliterated more national forest roads than they built (see Figure 20). Consequently, since 1991 roadless areas in the Northwest state's national forests have stopped shrinking, the first such pause in road expansion in generations. (In fact, further analysis might reveal that roadless areas—large expanses of territory unbroken by roads—have actually begun expanding. Unfortunately, the data presented here do not speak to that question, because they do not reveal the locations of obliterated roads; moreover, data on BC's innumerable logging roads are simply unavailable.)[50]

The Northwest's roads are indicators of the physical scale of northwesterners' real-estate development and logging. They are also a first-order ecological problem in themselves. Roads fragment previously intact ecosystems, speed the spread of invasive organisms, cause landslides on steep slopes, and send pollution and sediment into nearby waterways.[51]

The building boom in US national forests peaked around 1980, when the Forest Service constructed (and reconstructed) about 5,000 miles of logging roads across the Pacific Northwest in a single year. In the early 1990s, the Forest Service officially shifted its emphasis from logging to ecosystem management, and road building fell to about one-tenth its former level. Construction of new roads was measured in tens,

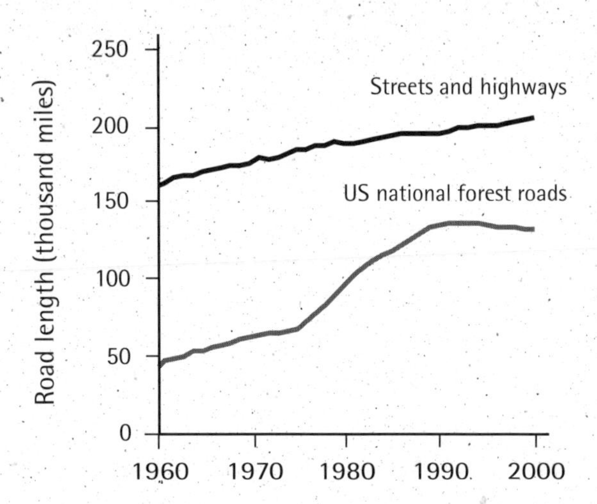

Figure 19. Throughout the Pacific Northwest, streets and highways continued to lengthen, but US national forest roads began to decline. Sources: see endnote 50.

not thousands, of miles. And the Forest Service began obliterating hundreds of miles of forest roads, recontouring and replanting them. Going further, the service blocked access to more roads than it obliterated.[52]

The US national forest road system in the Pacific Northwest, excluding northwestern California, shrank by 4,400 miles between 1991 and 2000 as obliteration outpaced new road construction; the Forest Service also closed an additional 17,200 miles of roads. Unlike obliterated roads, of course, closed roads can be reopened, and they remain an ecological problem. Though they may erode more than maintained roads, they do return to a more natural state over time.[53]

Information about the Northwest's road network is incomplete. The inventoried streets, highways, and Forest Service roads we chart here may account for only half of all roads that spread across the region. Unrecorded "ghost roads" are commonplace on national forest lands. BC's enormous public lands are laced with logging roads built by timber companies, which are not required to release to the public information on their full-throttle road construction. And lands managed by private timber owners, state agencies, the US Bureau of Land Management, and others hold tens of thousands more miles of roads. Together, the Northwest's road network likely stretches more than 800,000 miles, enough to circle the equator 32 times and roughly as far as all the region's streams taken together—meaning that cars and trucks may now have better access to the outdoors than do salmon.[54]

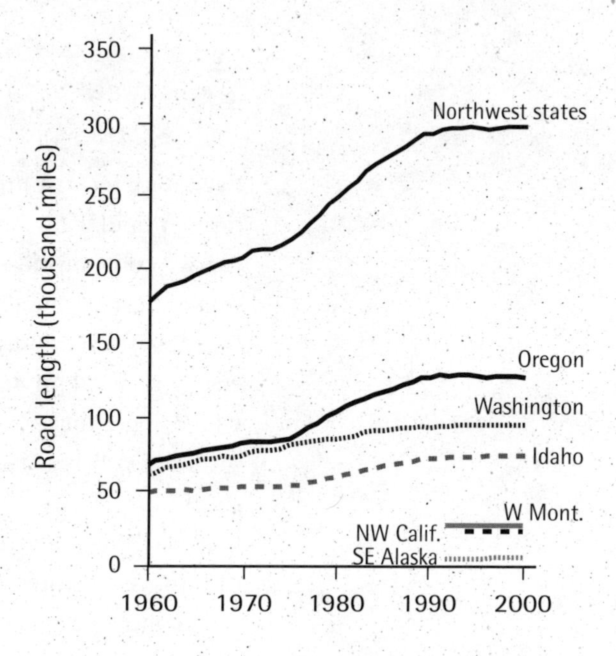

Figure 20. In the Northwest states, the pace of new habitat fragmentation slowed in the 1990s, as the road network stabilized. Sources: see endnote 50.

ENERGY

indicator

Even with modest improvements in energy efficiency during the 1990s, rapid population growth steadily pushed up the Northwest's total energy consumption (see Figure 21). By 2000, northwesterners consumed nearly a supertanker's worth of energy every two days. Still, the Northwest's economy is growing far faster than its energy use, as knowledge and human services, rather than energy-intensive industries, increasingly fuel the region's prosperity.[55]

Energy brings enormous benefits to northwesterners' lives but also substantial environmental and social costs, many of them hidden. Burning fossil fuels—petroleum products, natural gas, and coal—pollutes our air, contributes to global warming (see "Greenhouse Gases"), ties our economy to unstable commodity markets and distant oil fields, and threatens the continent's remaining wild areas. Hydropower dams, the Northwest's principal source of electricity, harm salmon runs.

Long accustomed to low electricity prices, northwesterners faced a sobering new reality in 2001: no longer can the region count on seemingly unlimited cheap power. Record low precipitation curtailed hydroelectric production early in the year, sending wholesale power prices soaring. In response, many Northwest utilities raised their rates and bolstered their conservation programs, hoping to keep California's power blackouts from rolling northward. The reaction by consumers was impressive: Seattle's electric utility, for example, reported that by late October, its customers had reduced their total electricity use for the year by 8 percent below forecast levels.[56]

These conservation gains may turn out to be short-lived, but they are consistent with a longer-term trend toward more efficient use of energy

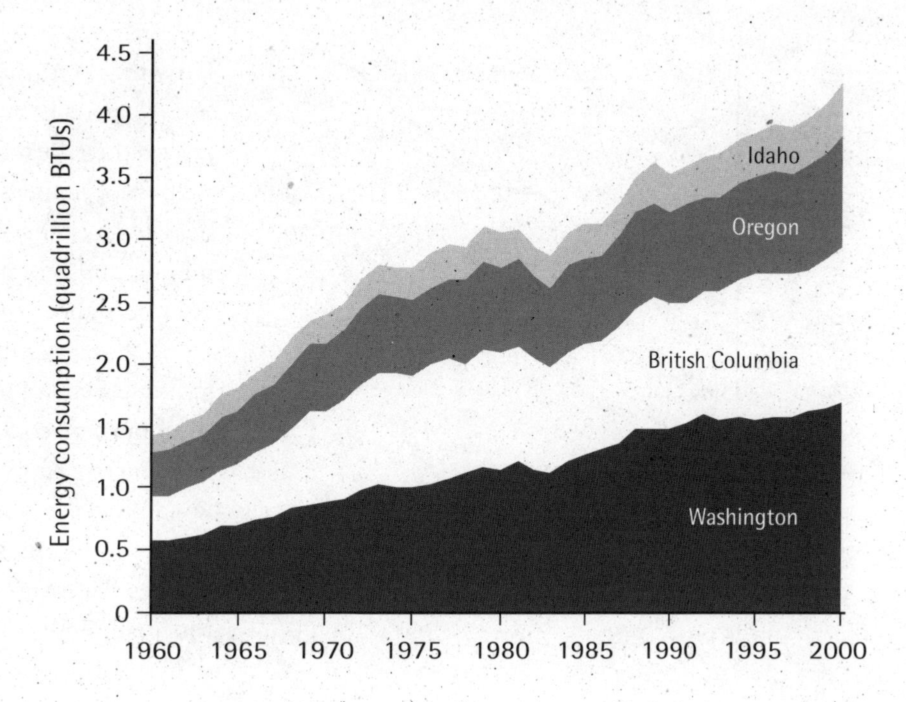

Figure 21. The Northwest's appetite for energy has tripled since 1960. Sources: see endnote 55.

in the region. In 1981 the economies of Washington, Oregon, and Idaho generated about $464 for every 5.8 million BTUs (British thermal units), the amount of energy in a barrel of oil. By 2000 they produced $712, adjusted for inflation (see Figure 22). This trend reflects the region's shift away from heavy industry: between 1978 and 2000, energy consumption by stores and office buildings in the Northwest increased three and a half times faster than industrial use. Industry's share of energy use declined further in 2001, with the closure of 10 of the region's 11 power-hungry aluminum smelters. Smelters in Washington and Oregon have been responsible for up to 16 percent of all electricity consumed in those states.[57]

Although BC and US Northwest residents use energy at similar rates, they use it in different ways. Compared with their neighbors to the south, British Columbians use about one-fourth less energy per person for cars, trucks, trains, and airplanes and about one-fifth less to heat and power their homes. These efficiencies likely result from smaller homes and more compact, transit-friendly urban design (see "Smart

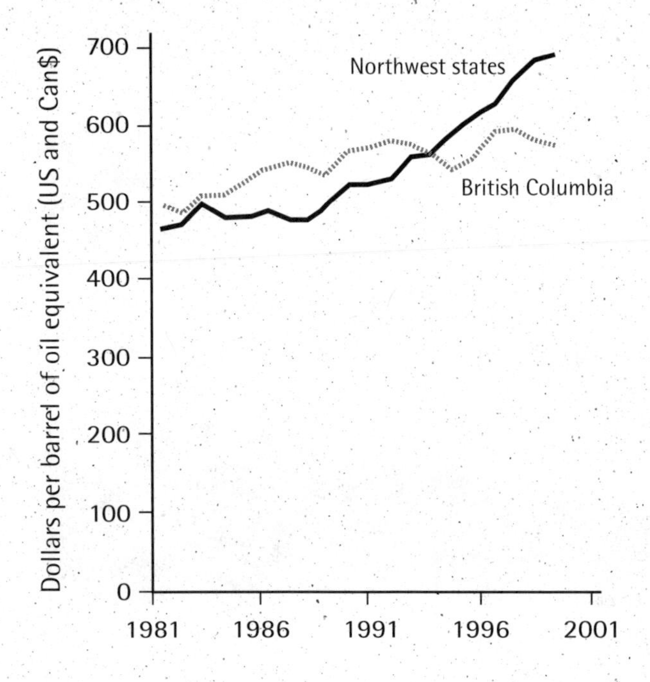

Figure 22. The Northwest is squeezing more goods and services from each unit of energy it consumes. Sources: see endnote 57.

Growth"). On the other hand, BC residents use 40 percent more energy per person to power their businesses and factories—a sign that the province's economy depends more heavily on energy-intensive industries, such as mining and paper mills, than do the economies of Idaho, Oregon, and Washington.[58]

Although hydropower still dominates, a growing share of the region's electricity comes from burning coal and natural gas. A sustainable energy supply would come more from wind power and other renewables and entail vastly more efficient use of present energy sources. The region might then trim its demand for fossil fuels or even export large quantities of hydropower, as it did until the late 1970s.[59]

A more comprehensive indicator of northwesterners' resource consumption would measure the region's entire diet of commodities extracted from farms, fisheries, forests, mines, rangelands, and wells. Best available estimates and national data suggest that northwesterners consume, on average, their body weight in such goods every day, but few time series are tabulated at state and provincial levels for resources besides energy.[60]

GREENHOUSE GASES

indicator

Despite substantial reductions in certain greenhouse gases, annual climate-altering emissions from the Northwest remained roughly constant in 2001. Early in the year, 10 of the region's 11 aluminum smelters closed, eliminating the source of 3 to 4 percent of the region's greenhouse gas emissions. Northwesterners' voracious appetite for fossil fuels, however, likely offset these declines, as increased consumption of motor gasoline and natural gas sent more carbon dioxide skyward.[61]

Greenhouse gas emissions are an important gauge of the region's well-being because global warming threatens both people and nature. Like other North Americans, northwesterners are responsible for a vastly disproportionate share of heat-trapping gases—more than three times the world average. At naturally occurring levels, these gases act like a layer of blankets, keeping the planet warm, humid, and hospitable. But billions of tons of emissions from cars, factories, and farms are piling on extra blankets. The planet's 12 warmest years on record have all come since 1980, and the 1990s were the warmest decade of the last millennium. Scientists expect warming trends to continue. The Northwest may lose some of its winter chill, and growing seasons at northern latitudes may lengthen. But other effects are more worrisome. Sea levels will rise, threatening coastal cities; croplands will turn to desert; snowpacks will shrink; glaciers will disappear; rivers will dry up, even as winter floods worsen; ecosystems will be transformed; and wildlife will be displaced.[62]

Smelters in Washington, Oregon, and Montana typically produce nearly 40 percent of the United States' aluminum and consume one-sixth of the region's electricity. Aluminum carries a high ecological toll:

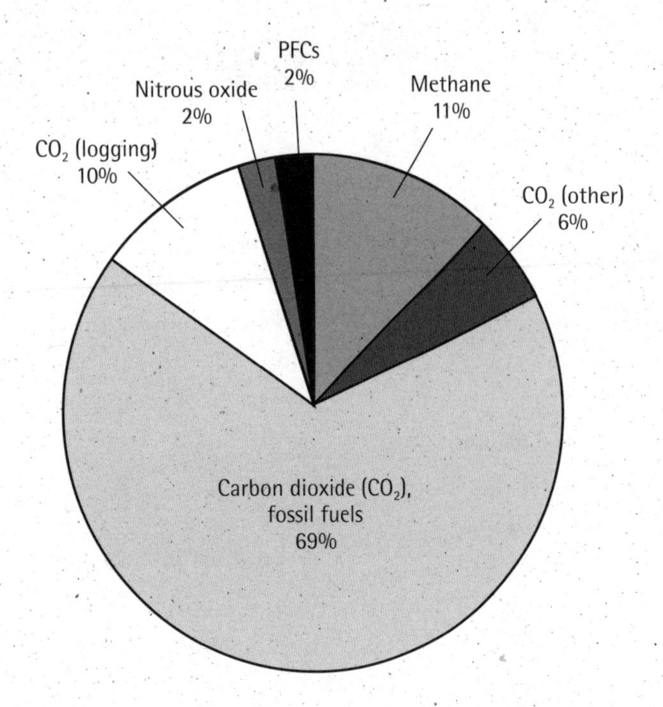

Figure 23. Fossil fuels are the Northwest's main greenhouse gas emitter. Sources: see endnote 65.

the smelters generate nearly twice as much climate-warming carbon dioxide as they do aluminum, and they discharge compounds known as perfluorocarbons (PFCs) into the sky. Molecule for molecule, PFCs trap heat more than 6,000 times as effectively as carbon dioxide, and they remain in the atmosphere for tens of thousands of years. In recent years, aluminum smelters have accounted for as much as 4 percent of the region's greenhouse gas tab.[63]

But in 2001, the ten aluminum smelters in Washington, Oregon, and Montana closed. California's power shortage, coupled with scarce rainfall, sent electricity rates soaring, and smelter operators found that they could earn higher profits selling their guaranteed cheap power back to the grid than by continuing to operate. Shutting down the smelters cost many workers their jobs. But by year's end, the smelter closures had slashed the region's PFC production to near zero and also modestly reduced emissions of carbon dioxide.[64]

Despite these reductions, the region realized little or no decline in total climate-altering emissions because the hydropower shortage that stalled aluminum smelters also increased the use of fossil fuels to generate electricity. Over the last two decades, swelling population, rising affluence, automobile-centered communities, and poor fuel efficiency have kept consumption mounting. Burning fossil fuels now accounts for nearly 70 percent of the Northwest's total contribution to global warming (see Figure 23). The region sent an estimated 194 million tons of carbon dioxide from fossil fuels into the sky in 2001, the daily equivalent of 70 pounds of carbon dioxide per northwesterner (see Figure 24). Fossil-fuel emissions may have diminished somewhat in the waning months of 2001, as the economy slowed and air travel decreased after September 11, but increased fossil-fuel electricity production, coupled with population-driven increases in petroleum consumption, likely outweighed these reductions.[65]

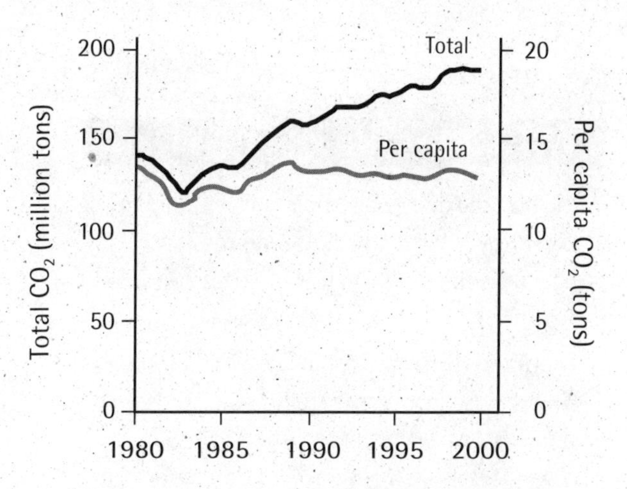

Figure 24. CO_2 emissions from fossil fuels have climbed 19 percent since 1990, but per capita emissions have stabilized. Sources: see endnote 65.

Other major sources of greenhouse gas emissions appear to have stabilized or declined slightly in 2001. Old-growth logging, the region's second-biggest source of carbon dioxide emissions, remained roughly steady. When the Northwest's ancient coastal forests are cut—and when their bark, waste wood, wood chips, sawdust, paper products, and carbon-rich forest soils decay or are burned—massive amounts of carbon dioxide return to the atmosphere. Logging these forests costs the region in another way: the forests act as "carbon sinks," soaking up carbon from the atmosphere, in effect serving as an emissions debit against its ever mounting credit bill. Virtually no old-growth trees were felled in 2001 in Oregon and Washington, where few ancient rainforests remain, and those few are largely protected in national parks and wilderness areas. But in British Columbia, the cut of old-growth forests continued aggressively.[66]

The region's methane emissions appear to be diminishing, at least somewhat. Landfill methane emissions, arising from garbage rotting in the absence of oxygen, are likely declining because solid-waste

What Goes Up

Carbon dioxide (CO_2), mainly from logging old-growth forests and burning fossil fuels, accounts for about 85 percent of the Northwest's contribution to climate change.

Methane (CH_4), the second most common greenhouse gas, comes from domesticated animals, especially cattle; waste disposal; and natural-gas leaks. Its heat-trapping potential is 21 times that of CO_2.

Nitrous oxide (N_2O) comes mainly from fertilizers and automobiles. Its heat-trapping potential is 30 times that of CO_2.

PFCs (perfluorcarbons) come from aluminum smelters. They occur nowhere in nature, and their heat-trapping potential is 6,000–9,000 times that of CO_2.

Sources: see endnote 68.

managers increasingly capture the methane and flare it off or sell it as a substitute for fossil fuels such as natural gas or coal. Methane emissions from cattle have stabilized, and cattle numbers have actually tapered off—a hopeful sign that livestock emissions may soon be on the decline as well.[67]

Though warming will stress the Northwest's natural systems, our prosperity enables us to better withstand the effects of climate change than most of the globe's inhabitants. Yet in 2001 a troubling picture unfolded. Emissions stabilized, but mostly because of drought and a slow economy rather than smart policies and conscious choices. And we missed a rare opportunity: perfluorocarbon production was virtually halted, but our net warming emissions did not fall. In an increasingly interdependent world, our ability to rein in oil, gas, and coal consumption will determine whether we are neighbors worth having.

MEASURING WHAT MATTERS

THREE MINOR CONCLUSIONS, AND ONE MAJOR one, emerge from the provisional indicators in *This Place on Earth 2002*. First, the current recession notwithstanding, human northwesterners—with the notable exception of the poor—are doing better than the nonhuman Northwest. Life expectancy trends are, if not as good as we could hope, certainly as good as we could expect. And income, though not equitably distributed, has mostly risen over recent years. But, as many of the indicators show, the region's natural estate is badly degraded and deteriorating annually.

Second, the car and its accoutrements of roads and auto-centered cityscapes are prime culprits in this degradation. From rising greenhouse gas emissions to stubbornly high energy consumption, from land-gobbling sprawl to habitat-fragmenting roads, the car—or, increasingly, the truck—lies at the root of many ills. It even diminishes life expectancy, since collisions kill so many northwesterners.

Third, now that per capita impacts have largely stabilized, population increases are behind much of the growth in environmental harm in the region. Still, environmental impacts remain exceptionally high per person in the Northwest, compared with global averages and historical precedents. Residents consume, on average, their body weight in natural resources each day, and northwesterners' expanding appetites have wiped out all the efficiency gains pouring out of research laboratories. Were it not for their ever larger homes, vehicles, and accumulations of consumer goods, northwesterners' environmental impacts would have been shrinking for years. So, to improve its long-term prospects dramatically, the region can apply itself to slowing population growth at the same time that it seeks to foster both efficiency and its necessary complement, a sense of sufficiency—an ethic of "enough is enough." (Tangible suggestions about pursuing these objectives can be found in Northwest Environment Watch's *This Place on Earth 2001*.)

The major conclusion, however, is that timely and trustworthy measures of the things that northwesterners cherish are remarkable mostly by their absence. As a result, the indicators in this book can hardly pretend to be comprehensive. They suggest a potential but do not realize it, because they do little more than begin to measure what matters—whether our communities and the natural systems they rely on are secure and thriving. The root of the problem is data: Canada and the United States track their finances assiduously but do not similarly monitor other spheres of life. The United States, for example, tracks bankruptcies, business starts, and money supply weekly, reporting the results in a matter of days. But it tracks child abuse, crime, income distribution, infant mortality, poverty, and teen suicide only annually and, even then, releases the data after many months' delay.[69]

Likewise, Oregon does not calculate the single best measure of human health—life expectancy—at all. British Columbia turns a blind eye to the logging roads endlessly partitioning and repartitioning its woodlands. The province does not survey the development of rural land, and the United States does so only at five-year intervals. Population density data, critical for planning walkable communities, come every five years in Canada and every ten in the United States. Moreover, not only for land development and population density but also for energy consumption, governments typically release the data when they are 18 months out of date.

To measure well what really matters—particularly things such as the status of the region's natural ecosystems and northwesterners' satisfaction with life—will require better data and further research, building upon the many indicators of progress developed since the 1992 Earth Summit in Rio de Janeiro. In the Pacific Northwest, the nonprofit group Sustainable Seattle's indicators are perhaps best known, but indicators projects have also sprung up at different times in small towns such as Vanderhoof, BC, and Sitka, Alaska; rural areas around bodies of water such as Willapa Bay, Washington, the lower Columbia River, and Flathead Lake, Montana; small cities such as Missoula, Montana, and Olympia, Washington; suburban areas such as Sonoma County, California; and even vast watersheds such as the Fraser River basin. Major metropolitan counties such as Oregon's Multnomah and Washington's King and Pierce have indicators of their own. British Columbia and Washington each have multiple sets of indicators in print. The region's state and provincial environmental agencies, along with their federal counterparts, have explored a regionwide set of environmental indicators. Oregon has a comprehensive state-supported system for monitoring progress, called Oregon Benchmarks, a biennial set of 90 indicators. This proliferation of projects is itself indicative of the region's hunger for a better barometer of progress.[70]

Measures of what northwesterners cherish are remarkable mostly by their absence

Filling the void would not be terribly difficult or expensive. The Northwest could develop a powerful battery of signals to assess its progress for perhaps $5 million a year—roughly what the region currently spends on soft drinks in a day.[71]

Here are a number of promising options:

We could develop a better set of signals for roughly what we spend on soft drinks in a day

- **Survey contentment.** The Northwest could monitor its residents' welfare by asking residents how happy they are. Northwesterners are polled and surveyed almost ceaselessly, but no one regularly asks them how satisfied they are with life in the region. Called "subjective well-being" by social psychologists, and measurable using established survey techniques, the "happiness quotient" reveals far more about human development than do economic data such as home size or disposable income. Appraising the region's satisfaction quarterly would widen public debate from its present blinkered gaze at finance to a full-viewed consideration of the quality of life. And, in a society where the vacillations of the Dow and the S & P 500 are reported ubiquitously and almost instantaneously, assessing human contentment quarterly seems only fair.[72]
- **Measure wealth, not just income.** The Northwest could assess, through annual or quarterly household surveys, the changing distribution of wealth—not income—in the region. Now measured infrequently, and only at the national level, wealth distribution is the single best gauge of how widely society shares the fruits of its prosperity, because wealth (or net worth) is a more important determinant of economic security than income. (And even income distribution is poorly monitored by the states and provinces, with results reported about two years late.) Regular reporting of wealth inequality would also focus poverty programs on the right question: how to enable have-nots to accumulate assets that appreciate in value. To date, the poor have been locked out of these capital gains.[73]

presidential approval rating   price of gold per ounce   price-to-earnings ratio   prime rate

Measuring wealth distribution is neither simple nor overwhelmingly complicated. Well-designed random-sample surveys would suffice, and the whole effort would not be much more expensive than the apparatus that currently reports housing starts each month in the Pacific Northwest. The wealth survey, in fact, would complement the data on housing starts, because it would tell the Northwest whom all those dwellings were likely for: second homes for the affluent or first homes for others.

- **Test for chemical "body burdens."** Beginning in 1999, the US government's Centers for Disease Control and Prevention began sampling human blood and urine from thousands of Americans for a battery of dangerous metals and synthetic chemicals. The results—an inventory of an American's body burden of toxins—constitute an excellent indicator of exposure to environmental pollutants and a biological counterpart to governments' monthly monitoring of personal income. The CDC's data show not what's coming into Americans' bank accounts but what's coming into their bodies. A similar annual census of toxins in northwesterners' bodies, perhaps adding breast milk to the samples collected, would allow the region to track its progress in detoxifying its human environment. No more expensive to gather than personal income data—since statistical methods and new, sensitive lab techniques allow relatively small samples to give accurate readings—tracking body burdens would also direct attention to those contaminants that are accumulating in the most worrisome proportions and in the most vulnerable groups of northwesterners, such as infants and children.[74]
- **Grade ecosystem health.** The Northwest has surprisingly little systematic information about the status of its natural capital—the communities of plants and animals that make human life possible and,

to many northwesterners, define their chosen land. Even the evolving status of the region's cultural icon and indicator animal, the salmon, remains—despite listings under the US Endangered Species Act and hundreds of millions of dollars spent in their aid—a great unknown in much of the region. At best the Northwest has snapshots of the abundance of some wild runs—usually those of commercial importance—from some time in the last decade. Vast stretches of salmon country in Alaska and British Columbia go unsurveyed for the fish. The dozens of distinct life histories that wild salmon follow in each watershed are disappearing even before they can be documented. And fisheries managers rarely monitor the underlying biological condition of the region's streams. Better, and regular, assessment of salmon status would serve as a rough proxy for the health of the ecosystems through which they travel, just as the status of other migratory species, such as songbirds or whales, indicates the status of the habitats they traverse.

Routine biomonitoring would emphasize the capacity of places to sustain life

But the Northwest could learn more about its natural capital if scientists—or even volunteers—across the Northwest collected sufficient data each year to fill in an index of biological integrity for each watershed. The index is a rigorous but simple report card on the status of ecosystems, which links the diversity, abundance, and ecological roles of different plants and animals to the amount and kind of disturbance a place has undergone. With appropriate adaptations, it is suitable for all types of landscapes and bodies of water. Such biomonitoring would put the emphasis in conservation where it belongs, on the capacity of places to sustain life.[75]

Comprehensive annual surveys either of indicator organisms such as salmon or of watersheds' indexes of biological integrity would not be cheap. But the region routinely surveys itself on a similar scale to monitor economic trends such as unemployment. Massive censuses of the work force lie behind the figures announced each month on joblessness, yet nature's work force—on which we ultimately depend not only for our livelihoods but for our lives—gets no similar scrutiny.

- **Map habitat fragmentation.** To improve on the partial measure of habitat destruction and fragmentation provided by the length of inventoried roads, the region could combine satellite imaging and aerial photography to scan the region's capillary network of roads as they branch and lengthen, and as they fade away. Similar techniques would allow annual reporting of changes in land use: the loss of natural forests, the contraction of streamside buffers, the shrinkage of roadless areas. Aquatic equivalents would map dams, dikes, levies, no-fishing zones, and shoreline development.

  Annual mapping of habitat fragmentation across the region would cost hundreds of thousands or even millions of dollars a year. But those sums are no more than various institutions spend gathering the data that make up the monthly index of leading economic indicators, which tries to predict impending shifts in broad economic trends. And the two are analogous: habitat fragmentation is a leading indicator of both species loss and the breakdown of ecosystem functions such as air and water filtration, nutrient cycling, and flood control.

- **Record consumption.** The national accounts that allow the calculation of gross domestic product and other macroeconomic indicators for the states and province of the Pacific Northwest—and the nations to which they belong—are a byzantine bookkeeping system that comprehensively tracks much of the money that flows through the economy. They are sophisticated enough to yield national GDP estimates on a quarterly basis after a lag of just a few weeks, along with reliable annual figures at the state and provincial level with a lag of just a few months.

But the Northwest boasts no comparable accounting system for the resources that flow through its economy, with the sole exception of energy. The region does not monitor its intake of food and fiber from farms, fisheries, and ranches; metals and other minerals from the Earth; paper and other wood products from forests; plastics and other synthetic materials from oil wells; or water from rivers and aquifers. Neither does anyone track the region's balance of trade in these things, as raw and finished goods cross in and out of the Northwest. Tracking resource flows along with financial flows would show the Northwest's environmental efficiency—and its inverse, wastefulness.[76]

- **Tally climate-changing emissions.** To better assess the global responsibility of its way of life, the region could tally its emissions of all greenhouse gases, improving on the estimates in this book. At present, no government agency publishes up-to-date state or provincial emissions figures even for CO_2 from fossil fuels—the quantities of which are simple to calculate from existing energy consumption data—much less for all greenhouse gases. Each month government reports on retail sales tell us what we spent filling our shopping carts, but we wait in vain for current reports on what we put in our atmosphere.
- **Count unplanned births.** Finally, as an indicator of the effectiveness of the region's family planning efforts—a sign of both the outlook for its children and whether it is taking responsibility for the global consequences of its population increases—the Northwest could monitor the share of its births that result from unintentional,

unwanted pregnancies. Washington already surveys mothers of newborns periodically on this question, and Alaska, Montana, and Oregon have committed to begin doing so. Following these states' lead, the region could conduct such surveys annually. A Northwest in which all children were planned and wanted would face far fewer tragedies of child abuse and abandonment; it would also grow more slowly.[77]

Indicators such as these are just a taste of what's possible; other ideas abound. As indicators of livable communities, the region could track what share of children can safely walk to a library or what share of residents trust their neighbors. To keep tabs on its natural capital, the Northwest could measure the "greenness"—strictly speaking, "net primary productivity," or photosynthetic output—of its aquatic and terrestrial ecosystems. Or it could regularly sample the organic matter in its soils—the accumulated nutrients that undergird its farms, forests, and grasslands.

If we do not measure what we value, we'll end up valuing what we measure

Donella Meadows, a pioneer in thinking about indicators before her untimely death in early 2001, collected a few other ideas for indicators of uncommon wisdom: How many hours of the day do people—and people of different classes, sexes, and races—have to work to meet their survival needs? Do children go on living in their communities after they grow up, or do they move away? Do people have to lock their houses and cars? What share of streams can you drink from safely? What share of people say they have enough? Do people smile at one another when they pass on the sidewalk?[78]

Skeptics about the utility of revising society's indicators could find no better spokesperson than the poet e. e. cummings. He wrote:

(While you and i have lips and voices which
are for kissing and to sing with
who cares if some oneeyed son of a bitch
invents an instrument to measure Spring with?

Why try to quantify the immeasurable, to apply calipers to our ultimate ends? Well, it turns out, there's an excellent reason. If we do not measure what we value, we'll end up valuing what we measure. That has happened to us. The Northwest is a region in the grips of one-eyed indicators. Lacking gauges for what we care about, we fall back on whatever signals we find: stock prices, the consumer confidence index, gross domestic product.[79]

And, reckoning our collective success by reference to those yardsticks, we organize our lives and institutions to mete out these things—to deliver rising stock prices, confident consumers, the grossest domestic product we can muster. Conversely, we do not get what we do not count: thriving, secure communities living amid vibrant, thriving nature. The point gives a new meaning to the biblical teaching, "The measure you give will be the measure you get."[80]

NOTES

1. Peter D. Ward and Donald Brownlee, *Rare Earth: Why Complex Life Is Uncommon in the Universe* (New York: Copernicus, 2000); and Donald Brownlee, Dept. of Astronomy, University of Washington, Seattle, private communication, Oct. 20, 2000.
2. More living matter from Edward O. Wilson, "Is Humanity Suicidal?" *New York Times Magazine,* May 30, 1993. Increased body size and lifespan from Robert William Fogel, *The Fourth Great Awakening and the Future of Egalitarianism* (Chicago: Univ. of Chicago Press, 2000). Population increase from US Census Bureau, *www.census.gov,* Dec. 4, 2000. Increased resource use from J. R. McNeill, *Something New under the Sun* (New York: Norton, 2000). Human influence on global climate from Intergovernmental Panel on Climate Change, *Climate Change 2001,* at *www.ipcc.ch*. Extinction rate from Edward O. Wilson, *The Diversity of Life* (Cambridge, Mass.: Harvard Univ. Press, 1992).
3. The Pacific Northwest comprises British Columbia, Idaho, Oregon, and Washington, along with Del Norte, Humboldt, Mendocino, Siskiyou, Sonoma, and Trinity Counties in California; Deer Lodge, Flathead, Granite, Lake, Lincoln, Mineral, Missoula, Powell, Ravalli, Sanders, and Silver Bow Counties in Montana; the Alaskan boroughs of Haines, Juneau, Ketchikan Gateway, Sitka, and Yakutat; and the Alaskan census areas of Prince of Wales–Outer Ketchikan, Skagway-Hoonah-Angoon, Valdez-Cordova, and Wrangell-Petersburg. In most cases, figures reported in this book for "the Northwest" exclude southeast Alaska, northern California, and western Montana, because data are hard to come by for the applicable jurisdictions alone. Northwest population from sources in note 24. Region's economy estimated from US Bureau of Economic Analysis (BEA), "Regional Accounts Data: Gross State Product Data," *www.bea.doc.gov/bea/regional/gsp,* and "National Accounts Data," *www.bea.doc.gov/bea/dn1.htm,* Nov. 26, 2001; and BC Stats, *www.bcstats.gov.bc.ca/DATA/BUS_STAT/bcea/bcgdp99.htm*. Intact ecosystems from John C. Ryan, *State of the Northwest,* revised ed. (Seattle: Northwest Environment Watch [NEW], 2000).
4. Eric Lipton, "Numbers Vary in Tallies of the Victims," *New York Times,* Oct. 25, 2001.

5. Walter Updegrave, "The Ins and Outs of Indexes," *Money,* Feb. 2000; Kenneth L. Fisher, "Never Say Dow," *Forbes,* Nov. 15, 1999; and Carol Vinzant, "The 'New' Dow Is Still a Relic," *Business 2.0,* Nov. 1999.
6. Dow Jones Indexes, "Dow Jones Averages," *www.djindexes.com/jsp/industrialaverages.jsp,* Dec. 5, 2001.
7. Clifford Cobb et al., "If GDP Is Up, Why Is America Down?" *Atlantic Monthly,* Oct. 1995; and Jonathan Rowe and Judith Silverstein, "The GDP Myth," *Washington Monthly,* March 1999.
8. Cobb et al., op. cit. note 7.
9. GDP per capita from US Census Bureau, *Statistical Abstract of the United States: 1991* and *1999* (Washington, DC: US Government Printing Office, 1991 and 1999). Share of Americans "very happy" from National Opinion Research Center, "General Social Survey Codebook," 1973–1998, University of Chicago, *www.icpsr.umich.edu/GSS,* Sept. 17, 2001; and Michael Worley, National Opinion Research Center, University of Chicago, private communication, Sept. 19, 1990. Kuznets quoted in Cobb et al., op. cit. note 7.
10. Inflation from Dean Baker, ed., *Getting Prices Right: The Debate over the Consumer Price Index* (New York: M. E. Sharpe, 1997); Norman Frumkin, *Tracking America's Economy,* 3d ed. (New York: M. E. Sharpe, 1998); and Advisory Commission to Study the Consumer Price Index, *Toward a More Accurate Measure of the Cost of Living [The Boskin Commission Report]* (Washington, DC: US Social Security Administration, 1996), at *www.ssa.gov/history/reports/boskinrpt.html.* Unemployment from US Bureau of Labor Statistics, "Technical Notes to Household Survey Data," *www.bls.gov/cps/cpstn1.htm;* and Economic Policy Institute, "A Glossary of Terms Used in State Data at a Glance," *www.epinet.org/datazone/states/usmap/glossary.html.* Poverty from Lee Rainwater and Timothy M. Smeeding, "Doing Poorly: The Real Income of American Children in a Comparative Perspective," Luxembourg Income Study Working Paper No. 127, Syracuse University, Syracuse, NY, 1995.
11. For thoughtful discussions of society's indicators, see Clifford W. Cobb and Craig Rixford, "Lessons Learned from the History of Social Indicators," Redefining Progress, San Francisco, 1998; Clifford W. Cobb, "Measurement Tools and the Quality of Life," Redefining Progress, San Francisco, 2000; and Donella Meadows, "Indicators and Information Systems for Sustainable Development," Sustainability Institute, Hartland Four Corners, Vt., at *www.sustainabilityinstitute.org,* Oct. 12, 2001.

12. Second homes in 1980 and 1990 from Myra Washington, Housing and Household Economic Statistics Division, US Census Bureau, private communication, Feb. 24, 1999; in 2000 from US Census Bureau, American FactFinder, "Housing Occupancy and Tenure (QT)," *factfinder.census.gov/servlet/BasicFactsServlet,* Aug. 2001. Number of second homes in 2000 not strictly comparable to previous years because of methodological differences in the housing censuses; see US Census Bureau, Housing Vacancies and Home Ownership, "Appendix A. Definitions and Explanations" and "Appendix B. Source and Accuracy of Estimates," *www.census.gov/hhes/www/housing/hvs/annual00/ann00ind.html.* Prison beds (prison inmates) from Bureau of Justice Statistics, *Sourcebook of Criminal Justice Statistics 1999* (Washington, DC: US Dept. of Justice, 2000), at *www.albany.edu/sourcebook;* and Allen J. Beck and Paige M. Harrison, "Prisoners in 2000," *Bureau of Justice Statistics Bulletin,* Aug. 2001, at *www.ojp.usdoj.gov/bjs/pub/pdf/p00.pdf.* Figure 1 excludes southeast Alaska, northern California, and western Montana.
13. See "Population," "Energy," and "Greenhouse Gases"; Figure 2 based on sources for these chapters. In Figure 2, trends for personal income, motor vehicles, and CO_2 emissions exclude southeast Alaska, northern California, and western Montana; road length excludes these Northwest states and BC; population includes the entire region.
14. Northwest greenhouse gas emissions goal from John C. Ryan, *Over Our Heads: A Local Look at Global Climate* (Seattle: NEW, 1997). Depletion of salmon runs from sources in note 38.
15. Life expectancy in BC from BC Stats, "British Columbia Life Expectancy," *www.bcstats.gov.bc.ca/data/pop/vital/bcexp.htm,* Oct. 17, 2001; in Idaho from Kathy Simplot, Bureau of Vital Records and Health Statistics, Idaho Dept. of Health and Welfare, private communication, July 11, 2001; and in Washington from Ann Lima, Center for Health Statistics, Washington Dept. of Health, private communication, July 10, 2001. The Oregon Health Division does not issue life tables: Joyce A. Grant-Worley, Center for Health Statistics, Oregon Public Health, private communication, July 6, 2000. Washington and Idaho 2001 increases in life expectancy estimated from rates for 1990 through 1999; BC 2001 increase estimated from rates for 1990 through 2000.
16. Annual increments from sources in note 15. US life expectancy from Arialdi Minino et al., "Deaths: Preliminary Data for 2000," *National Vital Statistics Reports,* Oct. 9, 2001, at *www.cdc.gov/nchs/data/nvsr/nvsr49/nvsr49_12.pdf.* Canadian life expectancy from US Central Intelligence Agency, "The World

Fact Book 2001: Life Expectancy at Birth," *www.cia.gov/cia/publications/factbook/fields/life_expectancy_at_birth.html,* Feb. 7, 2002.

17. Number of uninsured in the Northwest states estimated from Robert J. Mills, "Health Insurance Coverage: 2000," *Current Population Reports,* P60-215 (Washington, DC: US Census Bureau, 2001), at *www.census.gov/prod/2001pubs/p60-215.pdf.* The Washington State Population Survey puts the number of Washington residents without health insurance much lower than US Census Bureau estimates; see Washington Office of Financial Management (WOFM), *www.ofm.wa.gov/sps/2000/tabulations/cur_ins.htm.*
18. Disproportionate impact of young deaths derived from Life Expectancy, "The Life Table," *www.lifeexpectancy.com/lifetable.shtml,* Aug. 2001. Motor vehicle and cancer deaths in the Northwest states from Centers for Disease Control and Prevention (CDC), "CDC Wonder," *wonder.cdc.gov/mortsql.shtml,* Oct. 17, 2001 (motor vehicle deaths include ICD-9 classification codes 810.0 through 825.9, comprising both traffic and nontraffic deaths; traffic-related motor vehicle deaths account for 96 percent of all motor vehicle deaths from 1980 through 1999; cancer deaths include ICD-9 classification codes 140 through 208.9). Cancer incidence in Washington from Washington Dept. of Health, "Washington State Cancer Registry," *www3.doh.wa.gov/wscr/HTML/WSCRabout.shtm,* Aug. 2001.
19. Timothy M. Smeeding, "American Income Inequality in a Cross-National Perspective: Why Are We So Different?" Maxwell School of Citizenship and Public Affairs, Syracuse University, Syracuse, NY, April 1997; Health Alliance International and University of Washington International Health Program, "Health and Income Equity," *depts.washington.edu/eqhlth,* Oct. 17, 2001; Stephen Bezruchka, Dept. of Health Services, University of Washington, Seattle, private communication, Oct. 8, 2001.
20. Personal income for Washington, Oregon, and Idaho (1980–98 in 1996 dollars) from Washington State University Cooperative Extension, "Northwest Income Indicators Project," *niip.wsu.edu,* May 31, 2001. Personal income forecast (1996 dollars) in Washington from WOFM, "2001 Long-term Economic and Labor Force Forecast for Washington," *www.ofm.wa.gov/longterm/2001/longtermtoc.htm,* May 31, 2001; in Oregon from Oregon Office of Financial Analysis, "Top Ten Requested Data," *www.oea.das.state.or.us,* June 1, 2001; in Idaho from Idaho Division of Financial Management, "Idaho Annual Economic Forecast 2001," *www2.state.id.us/dfm/ief/2001/apr01/0104ief.html,* June 1, 2001. Poverty in Northwest states, 1980–99, derived

from US Census Bureau, "Historical Poverty Tables, Table 21," *www.census.gov/hhes/poverty/histpov/hstpov21.html,* Aug. 2001; 2000 derived from US Census Bureau, "State and County QuickFacts," *www.census.gov,* Aug. 2001; 2001 estimated as roughly equal to 2000. BC wealth distribution from Steve Kerstetter, "Behind the Numbers: BC Home to Greatest Wealth Gap in Canada," Canadian Centre for Policy Alternatives, *www.policyalternative.ca,* Nov. 22, 2001; and Steve Kerstetter, "BC's Bountiful Crop of Millionaires," Canadian Centre for Policy Alternatives, Aug. 13, 2001, at *www.policyalternatives.ca/bc/btn-bcmillionaires.pdf.* Number of BC poor from Statistics Canada (StatCan), "Table 202-0802: Persons with Low Income Before and After Tax, Showing Prevalence and Estimated Number, Annual," CANSIM II, *www.statcan.ca/english/CANSIM,* Oct. 24, 2001. Number of poor BC children from BC Campaign 2000, "Child Poverty in British Columbia: Report Card 2000," at *www.firstcallbc.org/content/poverty/report00/child_poverty_brochure.pdf,* Oct. 24, 2001. Figure 4 excludes BC, southeast Alaska, northern California, and western Montana.

21. Quintile averages estimated, using weighted averages based on middle-year contemporary populations, from Jared Bernstein et al., "Average Incomes of Fifth of Families in '78–'80 through '96–'98, by State," in *Pulling Apart: A State-by-State Analysis of Income Trends* (Washington, DC: Economic Policy Institute and Center on Budget and Policy Priorities, 2000), at *www.epinet.org/studies/pullingapart/1-18-00sfp.pdf.* Sport utility vehicle price from CarsDirect.com, manufacturer's suggested retail prices for 2002 Ford Expedition Eddie Bauer, 2002 Mercedes-Benz ML320 Base, and 2001 Land Rover Discovery Series II E, *www.carsdirect.com/home,* Nov. 21, 2001. Figure 5 excludes BC, southeast Alaska, northern California, and western Montana.
22. Households earning more than $100,000 a year from US Census Bureau, Census 2000 Supplementary Survey, State Profiles for Washington, Oregon, and Idaho, "Table 3. Profile of Selected Economic Characteristics," *www.census.gov/c2ss/www/Products/Profiles/2000/index.htm,* Sept. 2001. Households lacking assured access to adequate food at all times ("food insecure") from Mark Nord et al., *Prevalence of Food Insecurity and Hunger, by State, 1996–1998* (Washington, DC: US Dept. of Agriculture, 1999), at *www.ers.usda.gov/publications/fanrr2.* Second homes in 1980 from Myra Washington, op. cit. note 12; and in 2000 from US Census Bureau, American FactFinder, "Housing Occupancy and Tenure (QT)," *factfinder.census.gov/servlet/BasicFactsServlet,* Aug. 2001 ("seasonal, recreational, or occasional

use" and "other vacant" housing units). Overcrowded housing in 1980 from US Census Bureau, "Historical Census of Housing Tables: Crowding," *www.census.gov/hhes/www/housing/census/historic/crowding.html,* Aug. 2001; in 2000 from US Census Bureau, Census 2000 Supplementary Survey, "Profile of Selected Housing Characteristics," *factfinder.census.gov/home/en/c2ss.html,* Aug. 2001.

23. Alan Thein Durning, *Green-Collar Jobs* (Seattle: NEW, 1999), and Bernstein et al., op. cit. note 21.
24. Northwest population data, and data for Figure 6, from US Census Bureau and StatCan, as detailed below. NEW estimates of 2001 population growth based on annual increments in recent years, related trends such as employment rates, and midyear estimates from WOFM, Olympia; Oregon Center for Population and Census, Portland; and BC Stats, Victoria.

    Population of Idaho, Oregon, and Washington for 1900–89 from US Census Bureau, "Historical Annual Time Series of State Population Estimates and Demographic Components of Change: 1900 to 1990 Total Population Estimates," *www.census.gov/population/www/estimates/st_stts.html.* California, Montana, and Alaska decennial population for 1900–90 from US Census Bureau, *Population of Counties by Decennial Census: 1900 to 1990* (Washington, DC: US Census Bureau, 1995). Intercensal years interpolated from decennial figures and from California Dept. of Finance, "Intercensal Estimates of Total Population for California Counties, Censuses of 1940 and 1950, and Estimates 1947–1969"; "July Intercensal Population Estimates for California Counties, 1970–1980"; and "July Intercensal Population Estimates for California Counties, 1980–1990," California Dept. of Finance, Sacramento, n.d.

    Northwest states' population for 1990–99 from US Census Bureau, "ST-99-3, State Population Estimates: Annual Time Series, July 1, 1990, to July 1, 1999," *www.census.gov/population/estimates/state/st-99-3.txt,* Dec. 2001; and US Census Bureau, "CO-99-8, County Population Estimates and Demographic Components of Population Change: Annual Time Series, July 1, 1990, to July 1, 1999," *www.census.gov/population/estimates/county/co-99-8/99C8_01.txt,* Dec. 2001, adjusted in light of 2000 US Census from US Census Bureau, "State and County QuickFacts," *quickfacts.census.gov/qfd,* July 2001 (Idaho, California, Montana, and Alaska); WOFM, "2001 Population Trends for Washington State," *www.ofm.wa.gov/poptrends/poptrendtoc.htm,* June 2001 (Washington); and Oregon Office of Economic Analysis (OEA), *Oregon Economic and Revenue Forecast,* "Appendix C"

(Salem: Dept. of Administrative Services, 2001), at *www.oea.das.state.or.us/economic/forecast1201.pdf* (Oregon).

Estimates of BC population for census years 1901, 1911, 1921, 1931, 1941 from StatCan, "Historical Statistics of Canada, Section A. Population and Migration: A2–14, Population of Canada, by Province, Census Dates, 1851 to 1976," *www.statcan.ca/english/freepub/11-516-XIE/sectiona/sectiona.htm,* Dec. 2001 (intercensal years interpolated); for 1951 to 2001 from StatCan, "Table 051-0005: Estimates of Population, Canada, Provinces and Territories, Quarterly (Persons)," CANSIM II, *www.statcan.ca/english/CANSIM,* Dec. 2001.

25. Cost of growth from Patrick Mazza and Eben Fodor, "Taking Its Toll: The Hidden Costs of Sprawl in Washington State," Climate Solutions, Olympia, Feb. 2000, at *www.climatesolutions.org.* See also Alan Thein Durning and Christopher Crowther, *Misplaced Blame: The Real Roots of Population Growth* (Seattle: NEW, 1997).
26. Annual increments and population doubling dates based on sources in note 24. International growth rates and populations from *World Resources 2000–2001* (Washington, DC: World Resources Institute, 2000), at *www.wri.org.*
27. Birth and migration rates from US Census Bureau, "State Population Estimates and Demographic Components of Population Change: April 1, 1990, to July 1, 1999," *www.census.gov/population/estimates/state/st-99-2.txt,* Dec. 2001; StatCan, "Population," *www.statcan.ca/english/Pgdb/People/Population/demo02.htm,* Dec. 2001; StatCan, "Births and Birth Rate," *www.statcan.ca/english/Pgdb/People/Population/demo04a.htm,* Dec. 2001; StatCan, "Deaths and Death Rate," *www.statcan.ca/english/Pgdb/People/Population/demo07a.htm,* Dec. 2001; StatCan, *Annual Demographic Statistics,* Catalog 91-213 (Ottawa: StatCan, 1994); WOFM, "2001 Population Trends for Washington State," op. cit. note 24; and OEA, op. cit. note 24.
28. Share of births from unintended pregnancies from Stanley K. Henshaw, "Unintended Pregnancy in the United States," *Family Planning Perspectives,* Jan.–Feb. 1998; and M. E. Eaglin et al., *1996–1998 Washington State Pregnancy Risk Assessment Monitoring System (PRAMS) Surveillance Report,* vol. I (Olympia: Washington Dept. of Health, 2001), at *www.doh.wa.gov/cfh/PRAMS/Volume%20One%20complete%209_6_01.pdf.*
29. Durning and Crowther, op. cit. note 25.
30. Mapping and geographic analysis of population density by CommEn Space, *www.commenspace.org,* on the basis of data from the US Census Bureau

(Portland and Seattle metropolitan areas, 1990 and 2000) and StatCan, Census of Population (Greater Vancouver, 1986 and 1996).

31. Density thresholds from Peter W. G. Newman and Jeffrey R. Kenworthy, *Cities and Automobile Dependence* (Brookfield, Vt.: Gower Technical Press, 1989). These thresholds apply to urban cores and may not apply to smaller towns or isolated dense neighborhoods.
32. Alan Thein Durning, *The Car and the City* (Seattle: NEW, 1996); Robert Putnam, *Bowling Alone* (New York: Simon and Schuster, 2000); Chris Kochtitzky and Richard Jackson, "Creating a Healthy Environment: The Impact of the Built Environment on Public Health," Sprawl Watch Clearinghouse, Washington, DC, at *www.sprawlwatch.org*.
33. Impervious surface mapped by CommEn Space, *www.commenspace.org*, using Landsat TM and ETM imagery, with a combination of spectral-mixture analysis and interpretation of GIS data from the three metropolitan areas; change detection based on intercalibrated image pairs from 1988 and 1999 (Seattle), 1989 and 1999 (Portland), and 1987 and 1999 (Vancouver).
34. Impacts from 1000 Friends of Washington, "Land Use and Water Quality," *www.1000friends.org/waterq.htm*, Nov. 15, 2001; and US Environmental Protection Agency (USEPA), Office of Water, "Urbanization and Streams: Studies of Hydrologic Impacts," March 1998, at *www.epa.gov/OWOW/NPS/urbanize/report.html*.
35. Stream degradation at one house per acre from 1000 Friends of Washington, op. cit. note 34. Coho sensitivity and concentrated pavement from Tom Schueler, "The Importance of Imperviousness," *Watershed Protection Techniques*, fall 1994.
36. Population change for Portland and Seattle (1990–2000) and Vancouver (1986–96) from sources in note 30.
37. Salmon boom and bust in 2001 from US Army Corps of Engineers, Portland District, "Adult Fish Counts," *www.nwp.usace.army.mil/op/fishdata/Adultfishcounts.htm*, Nov. 26, 2001; Fish Passage Center, Portland, "Adult Data," *www.fpc.org/adult.html*, Nov. 26, 2001; Associated Press, "Good River Flows Several Years Ago Set the Stage for Record Fish Run," *Seattle Post-Intelligencer*, Aug. 29, 2001; Michele DeHart, "Spring Migration 2001," preliminary analysis from Fish Passage Center, Portland, Aug. 10, 2001, at *www.fpc.org/fpc_docs/200-01.pdf*; Jonathan Brinckman, "Fish Survival Rates Plunge to Near-Record Lows in 2001: Future Effects Unknown," *Oregonian*, Oct. 11, 2001; and Wesley Loy, "Many Western Alaska Fishermen Ready to

Quit, Survey Finds," *Anchorage Daily News,* Nov. 6, 2001. Climate cycles and salmon from James J. Anderson, "Decadal Climate Cycles and Declining Columbia River Salmon," in E. E. Knudsen et al., *Sustainable Fisheries Management: Pacific Salmon* (Boca Raton, Fla.: Lewis Publishers, 2000); Nathan J. Mantua et al., "A Pacific Interdecadal Climate Oscillation with Impacts on Salmon Production," *Bulletin of the American Meteorological Society,* June 1997; and Nathan J. Mantua, Joint Institute for the Study of the Atmosphere and Oceans, University of Washington, Seattle, private communication, Aug.–Sept. 2001.

38. Wild salmon status from, among others, Independent Scientific Group, "Scientific Issues in the Restoration of Salmonid Fishes in the Columbia River Ecosystem," *Fisheries,* March 1999; Independent Scientific Group, *Return to the River: Restoration of Salmonid Fishes in the Columbia River Ecosystem* (Portland: Northwest Power Planning Council, 2000), at *www.nwcouncil.org/library/return/2000-12.htm;* Robert T. Lackey, "Restoring Wild Salmon to the Pacific Northwest: Chasing an Illusion?" in Patricia Koss and Mike Katz, eds., *What We Don't Know about Pacific Northwest Fish Runs,* Portland State University, Portland, 2000; Jim Lichatowich, *Salmon without Rivers* (Washington, DC: Island Press, 1999); National Research Council (NRC), *Upstream: Salmon and Society in the Pacific Northwest* (Washington, DC: National Academy Press, 1996); Deanna J. Stouder et al., *Pacific Salmon and Their Ecosystems: Status and Future Options* (New York: Chapman and Hall, 1997); Charles F. Wilkinson, *Crossing the Next Meridian* (Washington, DC: Island Press, 1992); and Edward C. Wolf and Seth Zuckerman, eds., *Salmon Nation* (Ecotrust: Portland, 1999).
39. Ecological health from Robert M. Hughes, Dynamac, Corvallis, private communication, Aug. 2001; Ryan, op. cit. note 3; and Lichatowich, op. cit., note 38. Farm- and hatchery-raised fish from, among others, Bruce Barcott, "Aquaculture's Troubled Harvest," *Mother Jones,* Nov./Dec. 2001; and Ray J. White et al., "Better Roles for Fish Stocking in Aquatic Resource Management," *American Fisheries Society Symposium,* vol. 15, 1995.
40. C. Jeff Cederholm et al., *Pacific Salmon and Wildlife: Ecological Contexts, Relationships, and Implications for Management* (Olympia: Washington Dept. of Fish and Wildlife, 2000); G. V. Hildebrand et al., "Use of Stable Isotopes to Determine Diets of Living and Extinct Bears," *Canadian Journal of Zoology,* Nov. 1996; Robert Semeniuk, "Do Bears Fish in the Woods?" *Ecologist,* Dec. 2001; J. M. Helfield and R. J. Naiman, "Salmon and Alder as Nitrogen Sources

to Riparian Forests in a Boreal Alaskan Watershed," *Ecology,* Sept. 2001; Ted Gresh et al., "An Estimation of Historic and Current Levels of Salmon Production in the Northeast Pacific Ecosystem," *Fisheries,* Jan. 2000.

41. Current abundance from Lackey, op. cit. note 38; Lichatowich, op. cit. note 38; NRC, op. cit. note 38; Jim Lichatowich and Seth Zuckerman, "Muddied Waters, Muddled Thinking," in Wolf and Zuckerman, eds., op. cit. note 38; Thomas G. Northcote and Dana Y. Atagi, "Pacific Salmon Abundance Trends in the Fraser River Watershed Compared with Other British Columbia Systems," in Stouder et al., op. cit. note 38; and Fisheries Management Division, *Report of the Fraser River Panel to the Pacific Salmon Commission on the 1999 Fraser River Sockeye and Pink Salmon Fishing Season,* Pacific Salmon Commission, Vancouver, BC, March 2001, at *www.psc.org/Pubs/Frp99ar-screen.pdf.* Effects on people from, among others, Joseph Cone, *A Common Fate: Endangered Salmon and the People of the Pacific Northwest* (New York: Holt, 1995); Pacific Rivers Council, *A Call for a Comprehensive Watershed and Wild Fish Conservation Program in Eastern Oregon and Washington,* 2d ed., Pacific Rivers Council, Eugene; Charles F. Wilkinson, op. cit. note 38; and Wolf and Zuckerman, eds., op. cit. note 38.
42. Stock status by geography compiled by Dorie Brownell, Ecotrust, Portland, from T. T. Baker et al., "Status of Pacific Salmon and Steelhead Escapements in Southeast Alaska," *Fisheries,* Oct. 1996; C. W. Huntington et al., "A Survey of Healthy Native Stocks of Anadromous Salmonids in the Pacific Northwest and California," *Fisheries,* March 1996; W. Nehlsen et al., "Pacific Salmon at the Crossroads: Stocks at Risk from California, Oregon, Idaho, and Washington," *Fisheries,* March–April 1991; P. Higgins et al., "Factors in Northern California Threatening Stocks with Extinction," Humboldt Chapter, American Fisheries Society, Arcata, Calif., 1992; and T. L. Slaney et al., "Status of Anadromous Salmon and Trout in British Columbia and Yukon," *Fisheries,* Oct. 1996. Stock status compared with share of stocks evaluated from Jim Lichatowich, Alder Fork Consulting, Columbia City, Ore., private communication, Aug. 2001.
43. Daniel L. Bottom, Fish Ecology Division, National Marine Fisheries Service, Newport, Ore., private communication, Oct. 10, 2001; Daniel L. Bottom et al., *Salmon at River's End: The Role of the Estuary in the Decline and Recovery of Columbia River Salmon* (Seattle: National Marine Fisheries Service, 2001); Independent Scientific Group, *Return to the River,* op. cit. note 38; Phillip S. Levin and Michael H. Schiewe, "Preserving Salmon Biodiversity,"

American Scientist, May–June 2001; Jim Lichatowich, Alder Fork Consulting, Columbia City, Ore., private communication, Aug. 2001; Lichatowich, op. cit. note 38; James A. Lichatowich and Lars E. Mobrand, "Analysis of Chinook Salmon in the Columbia River from an Ecosystem Perspective," in *Applied Ecosystem Analysis: Background* (Portland: Bonneville Power Administration, 1995); and other sources in note 38.

44. James R. Karr, Aquatic and Fishery Sciences, University of Washington, Seattle, private communication, Dec. 2001; and Lichatowich, op. cit. note 38.

45. Figure 16 excludes southeast Alaska, northwestern California, and western Montana, and Figure 17 excludes these states and BC. Motor vehicle fleet in the Northwest (British Columbia, Idaho, Oregon, and Washington) includes passenger cars; light and heavy trucks; and all other commercial, government, and private vehicles intended for roadway use except motorcycles, golf carts, trailers, and farm vehicles. Vehicles in Washington, Oregon, and Idaho for 1900–99 from Office of Highway Policy Information (OHPI), "Section II: Motor Vehicles," *Highway Statistics Summary to 1995* and *Highway Statistics 1996* to *1999* (Washington, DC: US Federal Highway Administration [FHWA], 1996–99), at *www.fhwa.dot.gov/ohim/qfvehicles.htm.* Vehicles in 2000 for Oregon calculated on the basis of growth rates and data from Renee Davis, Oregon Dept. of Transportation, Salem, private communication, July 2001; for Idaho from Idaho Transportation Dept., Economics and Research Section, *www2.state.id.us/itd/econ/econpage.htm#Vehicle%20Registration %20Information,* July 2001; in 2000–01 for Washington from Judy Spencer, Dept. of Licensing, private communication, July and Aug. 2001.

BC vehicles for 1900–75 from F. H. Leacy, ed., *Historical Statistics of Canada,* 2d ed. (Ottawa: Statistics Canada, 1983); for 1976–98 from StatCan, "Table 405-0001: Road Motor Vehicle, Trailer, and Snowmobile Registration, Annual (Registrations)," CANSIM II, *www.statcan.ca/english/CANSIM;* for 1999–2000 from BC Stats, "British Columbia Licensed Passenger Vehicles as at December 31" and "British Columbia Licensed Commercial Vehicles as at December 31," at *www.bcstats.gov.bc.ca/data/dd/handout/mvlic.pdf.* Passenger capacity calculated on the basis of four seats with seat belts per motor vehicle in the Northwest, for a total of 48 million seat belts—nearly enough to carry 16 million northwesterners and 33 million additional Californians.

46. Vehicles from sources in note 45. Population from sources in note 24. Licensed drivers in Washington, Oregon, and Idaho from OHPI, "Section III: Driver Licensing," *Highway Statistics Summary to 1995* and *Highway Statis-*

tics 1996 to *1999* (Washington, DC: FHWA, 1996–99), at *www.fhwa.dot.gov/ohim/qfdrivers.htm;* in British Columbia from Paul Hardy, Corporate Research, Insurance Corporation of British Columbia, Victoria, private communication, Dec. 11, 2001.

47. Vehicles per capita calculated from sources in notes 24 and 45.
48. Motor vehicle costs and benefits from Durning, op. cit. note 32. Greenhouse gas emissions from Ryan, op. cit. note 14. Worst traffic in North America from Natalie Pawelski, "Study Finds Traffic Getting Worse," CNN.com, *www.cnn.com/2001/US/05/07/traffic.cities,* May 7, 2001. Motor vehicle accident deaths from CDC, op. cit. note 18.
49. Sources in note 45.
50. In Figure 19, "streets and highways" include BC but exclude southeast Alaska, northwestern California, and western Montana; "US national forest roads" also exclude these states. Street, road, and highway length in Northwest states from OHPI, *Highway Statistics 1960* to *1979* (Springfield, Va.: National Technical Information Service, 1961–80), available through FHWA at *www.fhwa.dot.gov/ohim/ohimstat.htm;* and OHPI, "Section V: Roadway Characteristics and Performance," *Highway Statistics Summary to 1995* and *Highway Statistics 1996* to *1999* (Washington, DC: FHWA, 1996–99), at *www.fhwa.dot.gov/ohim/qfroad.htm;* in southeast Alaska, northwestern California, and western Montana (limited data for 1990s only) from Rick Rogne, Montana Dept. of Transportation, Helena, private communication, Oct. 27, 1995; Michael Vigue, Alaska Dept. of Transportation, Juneau, private communication, Oct. 27, 1995; California Dept. of Transportation, *Assembly of Statistical Reports* (Sacramento: 1978–98). Oregon's highway length for 1980 and 1997 interpolated; 2000 roadway length estimated for each state.

    BC provincial street, road, and highway length from StatCan, *Road and Street Mileage and Expenditure* (Ottawa: StatCan, 1960–73); BC Ministry of Transportation (BCMoT), *Annual Report 1977–78* to *1993–94* (Victoria: BCMoT [former Ministry of Transportation and Highways], 1978 to 1994); and, for 1995–2001, Debra Crozier-Smith, Communications Branch, BCMoT, Victoria, BC, private communication, July 24, 2001. Data for 1994 and 1998 interpolated, 2000 estimated. BC municipal street, road, and highway length from Dan Carsen, Municipal Financial Services Branch, Ministry of Municipal Affairs, Victoria, private communication, Nov. 9, 1995; Municipal Financial Services Branch, yearly *Municipal Statistics* series, schedule 3 (Victoria: BC Ministry of Municipal Affairs), at *www.marh.gov.bc.ca/MUNFIN,* July

12, 2001; and Neil Goldie, Municipal Financial Services Branch, Ministry of Community, Aboriginal and Women's Services (former Ministry of Municipal Affairs), Victoria, private communication, July 13, 2001. Data for 1999 and 2000 are estimated.

Because BC forest roads are largely unmonitored—the provincial government's own forest roads are tabulated, but the vastly more extensive network built on provincial land by timber companies is not—BC is excluded from the discussion of forest roads. US national forest road length for 1991–2000 in Washington, 1989–2000 for Idaho and Oregon, 1988–2000 for western Montana, and 1987–2000 for Alaska calculated on the basis of US Forest Service (USFS), "FS Road Miles by Operational Maintenance Levels, Oct. 2000" (spreadsheet) from Jim Padgett, USFS, Washington, DC, private communication, Aug. 2001. Data for earlier years, interpolated where needed, from, among others, *Report of the Forest Service/Report of the Chief* (Washington, DC: USFS, 1960–94, 1996); "Annual Road Accomplishment Report" or "Final Accomplishment Report," 1994–98, for USFS Region 6 and Boise, Caribou, Payette, Salmon-Challis, Sawtooth, and Targhee National Forests; Paul Anderson, USFS Region 6, Portland, private communication, July 19, 2001; Donna Sheehy, USFS, Missoula, private communication, Oct. 13, 1995, and Aug. 2001; and Greg Watkins, USFS, Pleasant Hill, Calif., private communication, Oct. 11, 1995. All told, NEW gathered data on all national forests in Washington and Oregon and the following other national forests: Chugach and Tongass (Alaska); Bitterroot, Flathead, Kootenai, and Lolo (Montana); Boise, Caribou, Clearwater, Idaho Panhandle (includes Coeur d'Alene, Kanisku, and Saint Joe National Forests), Nez Perce, Payette, Salmon-Challis, Sawtooth, and Targhee (Idaho); and Klamath, Shasta-Trinity, and Six Rivers (California). For complete methods, contact NEW, Seattle.

51. *The Uncounted Costs of Logging* (Washington, DC: Wilderness Society, 1989); Reed F. Noss and Allen Y. Cooperrider, *Saving Nature's Legacy: Protecting and Restoring Biodiversity* (Washington, DC: Island Press, 1994); Schueler, op. cit. note 35; and Todd Litman, *Transportation Cost Analysis* (Victoria: Victoria Transport Policy Institute, 1995).
52. Obliteration, closure, and construction from forest road sources in note 50.
53. Problems of closed roads from Noss and Cooperrider, op. cit. note 51; and Keith J. Hammer, *The Road-Ripper's Guide to the National Forests* (Missoula: Watershed Center for Protection and Restoration, 1995).
54. John C. Ryan, "Roads Take Toll on Salmon, Grizzlies, Taxpayers," *NEW*

Indicator, Dec. 1995, at *www.northwestwatch.org/pubs/indic4.html;* and updated data in Ryan, op. cit. note 3.

55. Figure 21 excludes southeast Alaska, northern California, and western Montana. Washington, Oregon, and Idaho's energy consumption in 1960–99 derived from the Energy Information Administration's (EIA) State Energy Data System, *www.eia.doe.gov/emeu/sedr/contents.html,* July 19, 2001. For 2000, electric power sales estimated from Office of Coal, Nuclear, Electric, and Alternative Fuels, Tables 11, 47, and 65 in *Electric Power Monthly: March 2001* (Washington, DC: EIA, 2001), at *tonto.eia.doe.gov/FTPROOT/electricity/epm/02260103.pdf;* motor gasoline sales from FHWA, OHPI, "Monthly Motor Fuel Reported by States," Dec. 2000, *www.fhwa.dot.gov/ohim/mmfr/mmfrpage.htm,* July 2001; coal consumption from Office of Coal, Nuclear, Electric, and Alternative Fuels, Tables 38, 39, 42, and 44 in *Quarterly Coal Report: October–December 2000* (Washington, DC: EIA, 2001), at *tonto.eia.doe.gov/FTPROOT/coal/qcr/0121004q.pdf;* natural gas consumption from Office of Oil and Gas, Tables 15, 16, 17, 18, and 19 in *Natural Gas Monthly: June 2001,* at *tonto.eia.doe.gov/FTPROOT/natgas/ngm/01300106.pdf,* and Roy Kass, EIA, private communication, July 31, 2001; petroleum consumption on the basis of change in "product supplied" for Petroleum Administration for Defense District V, comprising Washington, Oregon, California, Arizona, Nevada, Alaska, and Hawaii, from Table 12 in *Petroleum Supply Annual 2000,* vol. 1, at *www.eia.doe.gov/pub/oil_gas/petroleum/data_publications/petroleum_supply_annual/psa_volume1/historical/2000/psa_volume1_2000 .html.* BC's energy consumption in 1980–99 from StatCan, "Table 128-0002: Supply and Demand of Primary and Secondary Energy in Terajoules, Quarterly," CANSIM II, *www.statcan.ca/english/CANSIM,* Aug. 17, 2001; data for 2000 estimated from the year's first two quarters.

    Energy consumption figures adjusted on the basis of the following factors: thermal generating plants supplying electricity to Idaho, Oregon, and Washington but located outside the region, and the share of PacifiCorp electricity sales from thermal generation outside the region, from Jim Lazar, consulting economist, Olympia, personal communication, Oct. 18, 2001; thermal conversion factors for fossil-fueled steam-electric plants from "Appendix C: Thermal Conversion Factors," *State Energy Data Report 1999* (Washington, DC: EIA, 2001), at *www.eia.doe.gov/emeu/sedr/contents.html;* petroleum energy consumption excluding lubricants, asphalt, road oil, and chemical feedstocks; and electricity transmission and distribution losses estimated as 10 percent of total electricity consumption. For a complete explanation of calculations, contact NEW.

BTUs per supertanker from Office of Industrial Technologies, Appendix F in *Energy, Environmental, and Economics (E3) Handbook* (Washington, DC: US Dept. of Energy, 1997), at *www.oit.doe.gov/E3handbook/appenf.shtml.*

56. Curtailed hydroelectric production from Office of Coal, Nuclear, Electric, and Alternative Fuels, Table 11 in *Electric Power Monthly: October 2001* (Washington, DC: EIA, 2001), at *tonto.eia.doe.gov/FTPROOT/electricity/epm/02260110.pdf.* Conservation by Northwest utilities from Seattle City Light, "Charts: Conservation," *www.ci.seattle.wa.us/light/ctracks.html,* Oct. 25, 2001; and Northwest Power Pool, "Weekly Report: Aggregate Historical Data," at *www.nwpp.org/weekly/historical_data.pdf.*
57. Figure 22 excludes southeast Alaska, northern California, and western Montana. Washington, Idaho, and Oregon economies (gross state product, 1996 dollars) for 1981–99 from BEA, op. cit. note 3, June 8, 2001; for 2000 from BEA, "Regional Accounts Data: Annual State Personal Income," *www.bea.doc.gov/bea/regional/spi,* June 4, 2001. Gross provincial product for British Columbia from BC Stats, "BC GDP at Market Prices and Final Domestic Demand 1981–1999," *www.bcstats.gov.bc.ca/data/bus_stat/bcea/bcgdp99.htm,* Sept. 2001. Smelters from Ryan, op. cit. note 14.
58. Regional differences in energy use from sources in note 55.
59. Shifting energy diet and hydropower exporting from sources in note 55.
60. Daily resource consumption from John C. Ryan and Alan Thein Durning, *Stuff: The Secret Lives of Everyday Things* (Seattle: NEW, 1997).
61. Smelter closures from Anonymous, "New Owner Shuts Longview Smelter," *Puget Sound Business Journal,* March 1, 2001; Gail Kinsey Hill, "Aluminum Industry Powering Down," *Oregonian,* March 11, 2001; Associated Press, "Aluminum Plant Workers in Limbo," *Tacoma Herald,* Aug. 10, 2001; and Gail Kinsey Hill, "Power Crisis Creates Turmoil for Troutdale Alcoa Employees," *Oregonian,* Aug. 12, 2001.
62. Ryan, op. cit. note 14; and Leonie Haimson, "How's the Weather? Taking the Earth's Temperature for 2000," *Grist Magazine,* Jan. 11, 2001, at *www.gristmagazine.com/grist/heatbeat/weather011101.stm.*
63. Smelters' aluminum production, electricity consumption, and PFC emissions estimated from Ryan, op. cit. note 14; PFC estimates assume 10 percent reduction in smelters' emission rates over the 1990s. Smelters' CO_2 emissions (1990) from John C. Ryan, "Greenhouse Gases on the Rise in the Northwest," *NEW*

Indicator, Aug. 1995, *www.northwestwatch.org/pubs/indic3 .html.* PFCs' heat-trapping potential from USEPA, "Global Warming Potentials," *www.epa.gov/globalwarming/emissions/national/gwp.html,* Oct. 14, 2001.

64. Sources in note 61; and Gail Kinsey Hill, "Power Pitfalls Include Hard Luck," *Oregonian,* May 5, 2001.
65. Fossil fuel's emissions share from Ryan, op. cit. note 14. Energy consumption for 1980–2000, op. cit. note 55; for 2001, derived from EIA, "Official Energy Statistics from the US Government: By State, 'Petroleum Product Sales' and 'Natural Gas Production and Sales'"; and EIA, "Official Energy Statistics from the US Government: By Fuel, 'US Natural Gas State Data' and 'US Petroleum State Data,'" *www.eia.doe.gov*, Oct. 30, 2001. CO_2 emissions coefficients for fossil fuels from "Comparison of EPA State Inventory Summaries and State-Authored Inventories," USEPA, at *yosemite.epa.gov/globalwarming/ghg.nsf/resources/PDFs/$file/pdfB-comparison1.pdf.* Figures 23 and 24 exclude southeast Alaska, northern California, and western Montana.
66. CO_2 from regional old-growth logging derived from Phil Comeau, Research Branch, British Columbia Ministry of Forests (BCMoF), Victoria, private communication, July 19, 1995; William Ferrell, Dept. of Forest Science, Oregon State University, Corvallis, private communication, Aug. 11, 1995; Ian Graeme, Analysis Section, BCMoF, Victoria, private communication, March 21, 2000; Bill Howard, Revenue Board, BCMoF, Victoria, private communication, March 29, 2000; BCMoF, "Table C-3: Volume of All Products Billed in 1999/00, by Species, by Forest Region," *Annual Report of the Ministry of Forests 1999/00* (Victoria: BCMoF, 2000), at *www.for.gov.bc.ca/pab/publctns/an_rpts/9900/AR1999_2000np.pdf;* BCMoF, "Table C-1: Volume of All Products Billed, by Species, by Forest Region," *Annual Performance Report of the Ministry of Forests 2000/01* (Victoria: BCMoF, 2001), at *www.for.gov.bc.ca/pab/publctns/an_rpts/0001/index.htm;* and Michael Milstein, "US Orders End to Old Growth Harvest," *Oregonian,* Jan. 9, 2001. NEW's estimates make the following assumptions: that old growth constitutes 98 percent of total timber harvested in BC; that harvesting of coastal old growth, not inland forests, releases the vast majority of CO_2 from logging in the Northwest; and that 25 percent of forest biomass is converted to CO_2 after logging. For a complete explanation of calculations, contact NEW.
67. Cattle in British Columbia from StatCan, "Table 003-0032: Number of Cattle on Farms by Class, Annual," CANSIM II, *www.statcan.ca/english/CANSIM/index.html;* in Washington, Oregon, and Idaho from US Dept. of Agriculture

(USDA), National Agricultural Statistics Service (NASS), "Published Estimates Data Base," *www.nass.usda.gov:81/ipedb,* Sept. 2001. Cattle methane emissions calculated on the basis of Office of Policy, Planning, and Evaluation, *State Workbook: Methodologies for Estimating Greenhouse Gas Emissions* (Washington, DC: USEPA, 1995); Oregon Dept. of Agriculture, Oregon Agricultural Statistics Service, "Oregon Agriculture and Fisheries Statistics: Cattle and Calf Inventory, 'Cattle and Calves: Number, Value, Cows and Calf Crop: Oregon, 1870–2000,'" at *www.oda.state.or.us/oass/bul0065.pdf;* USDA, NASS, "Washington Agri-Facts," Feb. 6, 2001, at *www.nass.usda.gov/wa/agri1feb.pdf;* and Idaho Agricultural Statistics Service (IASS), "Cattle and Calves: Inventory, by Classes and Weight, Idaho, January 1, 1992–01," *2001 Idaho Agricultural Statistics* (Boise: IASS and Idaho Dept. of Agriculture, 2001), at *www.nass.usda.gov/id/publications/annual%20bulletin/annbulltoc.htm.*

68. Greenhouse gas warming potentials from USEPA, "Global Warming Potentials," *www.epa.gov/globalwarming/emissions/national/gwp.html,* Oct. 18, 2001.
69. Frequency of measuring US indicators from Marc Miringoff and Marque-Luisa Miringoff, *The Social Health of the Nation: How America Is Really Doing* (New York: Oxford Univ. Press, 1999).
70. Northwest indicators projects as of Dec. 2001 include Sustainable Seattle, "Indicators of Sustainable Community," *www.scn.org/sustainable/indicat.htm;* Community Sustainability Auditing Project, *What Matters in Vanderhoof British Columbia?* at *web.uvic.ca/~csap/frbc/VAN.PDF;* Island Institute, *Sitka Community Indicators: A Profile of Community Well-Being* (Sitka: Island Institute, 1999), available through *home.gci.net/~island/CommunitySustainability;* Willapa Alliance and Ecotrust, "Willapa Indicators for a Sustainable Community: The 1998 Willapa Indicators," *65.165.109.4/wiscweb/WISC98.html;* Lower Columbia River Estuary Program, "Environmental Indicators," *www.lcrep.org;* Sunrift Center for Sustainable Communities, "Flathead Gauges: A Report to Citizens on Long-term Trends in Sustainability," Sunrift Center, Kalispell, Mont., 1997; Missoula County, "Missoula Measures," *www.co.missoula.mt.us/measures;* Sustainable Community Roundtable, "State of the Community" reports for South Puget Sound, at *www.olywa.net/roundtable;* Sustainable Sonoma County, *www.sustainablesonoma.org;* Fraser Basin Council, *Sustainability Indicators for the Fraser Basin: Workbook* (Vancouver: Fraser Basin Council, 2000), at *www.fraserbasin.bc.ca/documents/Indicator%20Workbook.pdf;* Portland Multnomah Progress Board, "Benchmark Areas," *www.p-m-benchmarks.org/*

tblcnts.html; King County Indicators Initiative, "Communities Count 2000," *www.communitiescount.org;* Pierce County Dept. of Community Services, "Quality of Life Benchmarks," *www.co.pierce.wa.us/services/family/benchmrk/qol.htm;* BC Ministry of Water, Land and Air Protection, State of Environment Reporting, "Environmental Trends in British Columbia 2000," *wlapwww.gov.bc.ca/soerpt/index.html;* University of Victoria, "Sustainable Communities Initiative," *web.uvic.ca/sci;* Washington State University, "Northwest Income Indicators Project: Washington," *niip.wsu.edu/washington/default.htm;* Environmental Health Programs, *Environmental Health Indicators* (Olympia: Washington Dept. of Health, 1998), at *www.doh.wa.gov/ehp/ts/EnvironmentalHealthIndicators.pdf;* Oregon Progress Board, "Oregon Benchmarks," *www.econ.state.or.us/opb.* See also Redefining Progress, "Community Indicators Projects on the Web," *www.rprogress.org/resources/cip/links/cips_web.html;* and International Institute for Sustainable Development, "Compendium of Sustainable Development Indicator Projects," *iisd1.iisd.ca/measure/compindex.asp.*

71. Northwest spending on soft drinks per day estimated from US and Canadian average per capita soft drink spending, Center for Science in the Public Interest, "Soft Drinks Undermining Americans' Health," Nov. 3, 1998, at *www.cspinet.org/new/soda_10_21_98.htm;* and from Agriculture and Agri-Food Canada, Food Bureau, "The Canadian Soft Drink Industry: SIC1111, Excluding Bottle Water Industry Data, 1988–1997," *www.agr.ca/food/profiles/softdrink/softdrink_e.html#Principle,* Nov. 21, 2001.
72. Measurement of happiness, or "subjective well-being," from Ruut Veenhoven, *Conditions of Happiness* (Dordrecht, Netherlands: D. Reidel, 1984); W. Pavot et al., "Further Validation of the Satisfaction with Life Scale: Evidence for the Cross-Method Convergence of Well-Being Measures," *Journal of Personality Assessment,* Aug. 1991; E. Sandvik et al., "Subjective Well-Being: The Convergence and Stability of Self-Report and Non-Self-Report Measures," *Journal of Personality,* Sept. 1993; and Ed Diener, "Assessing Subjective Well-Being: Progress and Opportunities," *Social Indicators Research,* Feb. 1994.
73. Edward N. Wolff, *Top Heavy: The Increasing Inequality of Wealth in America and What Can Be Done About It* (New York: New Press, 1996).
74. Body burden monitoring from National Center for Environmental Health, *National Report on Human Exposure to Environmental Chemicals* (Atlanta: CDC, 2001). Breast milk from Sandra Steingraber, *Having Faith: An Ecologist's Journey to Motherhood* (Cambridge, Mass.: Perseus, 2001).

75. Biomonitoring from James R. Karr, "Rivers as Sentinels: Using the Biology of Rivers to Guide Landscape Management," in R. J. Naiman and R. E. Bilby, eds., *River Ecology and Management: Lessons from the Pacific Coastal Ecoregion* (New York: Springer, 1998); James R. Karr and Ellen W. Chu, "Sustaining Living Rivers," *Hydrobiologia,* April 2000; John Whitfield, "Vital Signs," *Nature,* June 28, 2001; and Sarah A. Morley and James R. Karr, "Assessing and Restoring the Health of Urban Streams in the Puget Sound Basin," *Conservation Biology,* in press.
76. Materials accounting from, among others, Albert Adriaanse et al., *Resource Flows: The Material Basis of Industrial Economies* (Washington, DC: World Resources Institute and others, 1997); and Mathis Wackernagel and William Rees, *Our Ecological Footprint: Reducing Human Impact on the Earth* (Gabriola Island, BC: New Society, 1996).
77. Washington survey from Eaglin et al., op. cit. note 28. States participating in PRAMS from CDC, Reproductive Health Information Source, "Surveillance and Research," *www.cdc.gov/nccdphp/drh/srv_prams.htm,* Nov. 28, 2001.
78. Meadows, op. cit. note 11.
79. e. e. cummings, "voices to voices, lip to lip," *100 Selected Poems* (New York: Grove Press, 1989).
80. Mark 4:24, *New Oxford Annotated Bible with the Apocrypha,* 3d ed., New Revised Standard Version (New York: Oxford Univ. Press, 2001).